D1032237

volume 35

lecture notes in pure and applied mathematics

numbers of generators of ideals in local rings

Judith D. Sally

NUMBERS OF GENERATORS OF IDEALS IN LOCAL RINGS

PURE AND APPLIED MATHEMATICS

A Program of Monographs, Textbooks, and Lecture Notes

Contributions to *Lecture Notes in Pure and Applied Mathematics* are reproduced by direct photography of the author's typewritten manuscript. Potential authors are advised to submit preliminary manuscripts for review purposes. After acceptance, the author is responsible for preparing the final manuscript in camera-ready form, suitable for direct reproduction. Marcel Dekker, Inc. will furnish instructions to authors and special typing paper. Sample pages are reviewed and returned with our suggestions to assure quality control and the most attractive rendering of your manuscript. The publisher will also be happy to supervise and assist in all stages of the preparation of your camera-ready manuscript.

LECTURE NOTES
IN PURE AND APPLIED MATHEMATICS

1. *N. Jacobson,* Exceptional Lie Algebras
2. *L.-Å. Lindahl and F. Poulsen,* Thin Sets in Harmonic Analysis
3. *I. Satake,* Classification Theory of Semi-Simple Algebraic Groups
4. *F. Hirzebruch, W. D. Newmann, and S. S. Koh,* Differentiable Manifolds and Quadratic Forms
5. *I. Chavel,* Riemannian Symmetric Spaces of Rank One
6. *R. B. Burckel,* Characterization of C(X) Among Its Subalgebras
7. *B. R. McDonald, A. R. Magid, and K. C. Smith,* Ring Theory: Proceedings of the Oklahoma Conference
8. *Y.-T. Siu,* Techniques of Extension of Analytic Objects
9. *S. R. Caradus, W. E. Pfaffenberger, and B. Yood,* Calkin Algebras and Algebras of Operators on Banach Spaces
10. *E. O. Roxin, P.-T. Liu, and R. L. Sternberg,* Differential Games and Control Theory
11. *M. Orzech and C. Small,* The Brauer Group of Commutative Rings
12. *S. Thomeier,* Topology and Its Applications
13. *J. M. López and K. A. Ross,* Sidon Sets
14. *W. W. Comfort and S. Negrepontis,* Continuous Pseudometrics
15. *K. McKennon and J. M. Robertson,* Locally Convex Spaces
16. *M. Carmeli and S. Malin,* Representations of the Rotation and Lorentz Groups: An Introduction
17. *G. B. Seligman,* Rational Methods in Lie Algebras
18. *D. G. de Figueiredo,* Functional Analysis: Proceedings of the Brazilian Mathematical Society Symposium
19. *L. Cesari, R. Kannan, and J. D. Schuur,* Nonlinear Functional Analysis of Differential Equations: Proceedings of the Michigan State University Conference
20. *J. J. Schäffer,* Geometry of Spheres in Normed Spaces
21. *K. Yano and M. Kon,* Anti-Invariant Submanifolds
22. *W. V. Vasconcelos,* The Rings of Dimension Two
23. *R. E. Chandler,* Hausdorff Compactifications
24. *S. P. Franklin and B. V. S. Thomas,* Topology: Proceedings of the Memphis State University Conference
25. *S. K. Jain,* Ring Theory: Proceedings of the Ohio University Conference
26. *B. R. McDonald and R. A. Morris,* Ring Theory II: Proceedings of the Second Oklahoma Conference
27. *R. B. Mura and A. Rhemtulla,* Orderable Groups
28. *J. R. Graef,* Stability of Dynamical Systems: Theory and Applications
29. *H.-C. Wang,* Homogeneous Banach Algebras
30. *E. O. Roxin, P.-T. Liu, and R. L. Sternberg,* Differential Games and Control Theory II
31. *R. D. Porter,* Introduction to Fibre Bundles
32. *M. Altman,* Contractors and Contractor Directions Theory and Applications
33. *J. S. Golan,* Decomposition and Dimension in Module Categories
34. *G. Fairweather,* Finite Element Galerkin Methods for Differential Equations
35. *J. D. Sally,* Numbers of Generators of Ideals in Local Rings

Other Volumes in Preparation

NUMBERS OF GENERATORS OF IDEALS IN LOCAL RINGS

JUDITH D. SALLY
Department of Mathematics
Northwestern University
Evanston, Illinois

MARCEL DEKKER, INC. New York and Basel

Library of Congress Cataloging in Publication Data

Sally, Judith D.
Numbers of generators of ideals in local rings.

(Lecture notes in pure and applied mathematics)
Bibliography: p.
Includes index.
1. Local rings. 2. Ideals (Algebra)--Generators.
I. Title.
QA251.38.S24 512'.4 77-19016
ISBN 0-8247-6645-8

MARCEL DEKKER, INC.

270 Madison Avenue, New York, New York 10016

Current printing (last digit):
10 9 8 7 6 5 4 3 2 1

PRINTED IN THE UNITED STATES OF AMERICA

ISBN: 0-8247-6645-8

INTRODUCTION

In these notes we consider the problem of determining the number of generators of an ideal in a local (= commutative, Noetherian with unique maximal ideal) ring. Although there is an interplay of geometry and algebra in this problem, we look at recent (post 1965) results which use primarily algebraic techniques. These methods yield results sometimes in the form of bounds for numbers of generators and sometimes in the form of the ineligibility of certain numbers to be the number of generators of certain ideals.

The results which are covered in these notes can be viewed as very natural extensions of four classical theorems regarding numbers of generators. Three of these theorems concern bounds for numbers of generators.

1. Krull's principal ideal theorem (1928) gives a lower bound for the number of generators of an ideal I in a Noetherian ring R, namely, if I is of height n, then I needs at least n generators.

2. Macaulay [46] (1916) proved that there is no upper bound for the number of generators required for prime ideals in the polynomial ring $\mathbb{C}[X,Y,Z]$ and its localization $\mathbb{C}[X,Y,Z]_{(X,Y,Z)}$, where $\mathbb{C}$ is the field of complex numbers.

3. In 1938, Akizuki [4] proved that one dimensional local domains have a fixed bound on the number of generators of all ideals. In 1950, Cohen [22] gave a new proof of this fact and showed that the existence of such a bound characterizes one dimensional local domains.

The problem of determining when a variety is a complete intersection locally is the problem of determining when the bound given in Krull's theorem is attained, i.e., when can an ideal I in a local ring be generated by height I elements? Chapters 2 and 4 of these notes contain results on local complete intersections due to Davis [23], Ferrand [28], Szpiro [66] and Vasconcelos [68]. Chapter 2 also contains results due to Boratynski and Swiecicka [10], Eakin and Sathaye [26], Herzog and Waldi [34], Matlis [47], Northcott and Rees [52], Rees [18], Sally [59] and Singh [63] on the number of generators of powers of an ideal in a local ring.

In Chapter 3 we look at the question of how far the boundedness exhibited in Akizuki's theorem can be pushed. We also look at the open question of whether the unboundedness of Macaulay's primes is "typical" behavior for local rings of dimension three. Chapter 3 contains results due to Boratynski and Eisenbud [9], Kirby [40], Moh [50], Rees [56] and Sally [57] on questions involving boundedness in local rings of small dimension.

The boundedness results for Cohen-Macaulay local rings of small dimension follow from the fact that certain ideals in Cohen-Macaulay local rings of arbitrary dimension are bounded, where "certain" is often a property of the quotient ring. We will see in Chapters 2 and 5 that if R is a local ring and I an ideal, then properties of the quotient ring R/I sometimes give information about the number of generators of I. This kind of result has as its prototype the fourth classical theorem alluded to above:

4. Chevalley [21] (1943) proved that if $(R,\underline{m})$ is a regular local ring and I an ideal, then R/I is regular if and only if I is generated by a subset of a regular system of parameters for R (i.e., a subset of a minimal basis for $\underline{m}$.)

Chapter 5 contains results of Becker [8], Buchsbaum and Eisenbud [14], Hochster [8], Kunz [41], Sally [58] and

Watanabe [69] on the determination of the number of generators of I from the properties of R/I.

The main references we use are Bourbaki [11], Kaplansky [39], Matsumura [48], Nagata [51], Serre [61] and Zariski and Samuel [70]. To make these notes as self-contained as possible possible we have included in Chapter 1 statements and proofs of (almost) all the results used in the later chapters which are not found in one of the books mentioned above.

I wish to thank D. Eisenbud, M. Hochster, L. J. Ratliff, Jr. and W. Vasconcelos for helpful comments on the first draft of these notes. My thanks also go to V. Davis for the typing, to the National Science Foundation for support and to the Institute for Advanced Study for hospitality while these notes were being written.

TABLE OF CONTENTS

NUMBERS OF GENERATORS OF IDEALS IN LOCAL RINGS

CHAPTER 1

BASIC TOOLS

This chapter contains the fundamental tools needed for Chapters 2-5. In Section 0 we give some notation, terminology and abbreviations which will be used in the sequel. The reader is warned that the description of the "tools" given here is fashioned explicitly for use in these notes. However, full references are given so that the reader may look up the "generalization" of our "specialization."

0. Terminology, notation and abbreviations.

The terminology is most like [61]. The following list may help to avoid confusion where multiple names exist for certain concepts.

associated prime. The zero divisors, $\mathcal{Z}(A)$, on a finitely generated module module A over a Noetherian ring R are a finite union of primes: $\mathcal{Z}(A) = P_1 \cup P_2 \cup \cdots \cup P_t$, where $P_i = \text{ann}(a_i)$ for some nonzero element a_i in A. We use the terms P_i is an associated prime of A or P_i belongs to A. We write Ass $A = \{P_1, \ldots, P_t\}$.

If I is an ideal of R with irredundant primary decomposition $I = Q_1 \cap \cdots \cap Q_t$, where Q_i is primary for P_i, then we say that Q_i is the P_i-primary component of I.

depth. The depth of a finitely generated module A over a local ring $(R, \underline{m})$ is the length of a maximal A-sequence in $\underline{m}$. We avoid "grade" altogether and refer, when necessary, to the length of an A-sequence in the ideal I, if $I \neq \underline{m}$.

dimension. The dimension of a ring R is the Krull dimension; the dimension of an ideal I of R is the dimension of the ring R/I.

embedding dimension. The embedding dimension of a local ring $(R,\underline{m})$ is $\dim_{R/\underline{m}} \underline{m}/\underline{m}^2$, i.e., the number of elements in a minimal basis for $\underline{m}$.

local ring. A local ring is a Noetherian ring with unique maximal ideal.

minor of order t of a matrix M. A minor of order t of a matrix M is the determinant of a $t \times t$ submatrix of M.

quasi-local ring. A quasi-local ring is a ring with unique maximal ideal.

regular sequence.
A-sequence. If R is a Noetherian ring and A is a finitely generated R-module, then an A-sequence is a sequence $x_1, \ldots, x_s$ of elements of R such that $(x_1, \ldots, x_s)A \neq A$, and x_{i+1} is not a zero divisor on the module $A/(x_1, \ldots, x_i)A$, for $i = 0, \ldots, s-1$. When $R = A$ we will sometimes refer to an R-sequence as a regular sequence.

regular system of parameters. A regular system of parameters is a minimal generating set for the maximal ideal of a regular local ring.

ring. A ring is a commutative ring with identity.

Notation. k denotes a field.

$(R,\underline{m})$ denotes a local ring R with maximal ideal $\underline{m}$.

$\lambda_R(A) = \lambda(A)$ denotes the length of an R-module A.

$v(I)$ denotes the number of elements in a minimal basis of the ideal I in a local ring R.

$(B \underset{S}{:} A) = \{s \in S \mid sA \subseteq B\}$, where A and B are modules over a ring S and S is an R-algebra.

$(B \underset{R}{:} A) = (B : A)$

ann $A = (0 : A)$

[f] denotes the matrix with respect to fixed bases of a map f of finitely generated free R-modules.

Abbreviations. CM = Cohen-Macaulay
pd = projective dimension
s.o.p. = system of parameters.

1. Nakayama's Lemma.

The primary tool for counting generators in a local ring is, of course, Nakayama's lemma which reduces the problem of counting generators to that of counting vector space dimensions.

1.1 Nakayama's Lemma. Let $(R,\underline{m})$ be a local ring and A a finitely generated R-module. If $A = \underline{m}A$, then $A = 0$.

1.2. Corollary. Let I be an ideal in a local ring $(R,\underline{m})$. Elements $a_1,\ldots,a_n$ in I generate I if and only if the images of $a_1,\ldots,a_n$ in $I/\underline{m}I$ generate $I/\underline{m}I$ as a vector space over $R/\underline{m}$.

Remark. The corollary implies that the notion of minimal basis for I is well defined and that any set of generators of I contains a minimal basis.

Henceforth, we denote by $v(I)$ the number of elements in a minimal basis for the ideal I in a local ring $(R,\underline{m})$.

Note that if we make a change of rings $R \to R/L$, where L is any ideal of the local ring $(R,\underline{m})$ such that $L \cap I = LI$, then I and its image $(I,L)/L \cong I/LI$ have the same minimal number of generators, i.e., $v(I) = v((I,L)/L)$.

2. Hilbert functions.

Let A be an Artin ring and $G = \coprod_{\substack{n\in\mathbb{Z} \\ n\geq 0}} G_n$ a graded ring which is the quotient of a polynomial ring $A[X_1,\ldots,X_s]$ by a homogeneous ideal. Let $M = \coprod_{\substack{n \in \mathbb{Z} \\ n \geq 0}} M_n$ be a finitely generated graded module over G. Define a function on the nonnegative integers by $\Phi(M,n) = \lambda_A(M_n)$. In the case where A is a field, Hilbert proved that for large n, $\Phi(M,n)$ is a polynomial in n with rational coefficients of degree at most $s - 1$. Samuel later generalized the result to any Artin ring A.

Let $(R,\underline{m})$ be a d dimensional local ring and let I be an ideal of R. Hilbert functions are defined for I as follows.

For all nonnegative integers n,

$$H_{I,\underline{m}}(n) = \lambda(I^n/I^n\underline{m}).$$

If I is an $\underline{m}$-primary ideal, define

$$H_I(n) = \lambda(I^n/I^{n+1}).$$

It follows immediately from Hilbert's theorem that $H_{I,\underline{m}}(n)$ and $H_I(n)$ are polynomials for large n. In fact:

2.1. Hilbert-Samuel Theorem. For large n, $H_I(n)$ is a polynomial $P_I(n)$ in n of degree $d-1$ with coefficients in $\mathbb{Q}$, the rational numbers.

(A proof can be found in [48],[61] and [70], for example.)

We define the sum-transforms of the Hilbert functions as follows. Let I be an $\underline{m}$-primary ideal in the d dimensional local ring $(R,\underline{m})$.

$$H_I^0(n) = H_I(n)$$

and for positive integers j,

$$H_I^j(n) = \sum_{i=0}^{n} H_I^{j-1}(i).$$

The $H_I^j(n)$ are, for large n, polynomials with rational coefficients. In particular, $H_I^1(n)$ is, for large n, a polynomial $S_I(n)$ of degree d. We write $S_I(n)$ in the form

$$S_I(n) = a_d\binom{n+d}{d} + a_{d-1}\binom{n+d-1}{d-1} + \cdots + a_0,$$

where $a_d,\ldots,a_0$ are integers and $a_d \geq 1$. The multiplicity of I, $e(I)$, is defined by: $e(I) = a_d$.

When $I = \underline{m}$, we simplify the notation as follows: $H_{\underline{m}}^j(n) = H^j(n)$. $P_{\underline{m}}(n) = P(n)$, $e(\underline{m}) = e(R)$ and $e(R)$ is called the multiplicity of R.

Examples. (1). If $(R,\underline{m})$ is a d-dimensional regular local ring, then $H(n)$ is a polynomial for all $n \geq 0$. $H(n) = P(n) = \binom{n+d-1}{d-1}$ and $S(n) = \binom{n+d}{d}$. If I is an $\underline{m}$-primary ideal generated by an R-sequence of length d in a d-dimensional local ring $(R,\underline{m})$ then $H_I(n) = P_I(n) = \lambda(R/I)\binom{n+d-1}{d-1}$ and $S_I(n) = \lambda(R/I)\binom{n+d}{d}$.

(2). Let k be a field and $R = k[[x^3,x^4]]$. Then $\underline{m} = (x^3,x^4)$ and $\underline{m}^2 = (x^6,x^7,x^8)$, so that $\underline{m}^3 = x^3\underline{m}^2$. Thus $H(0) = 1$, $H(1) = 2$ and $H(n) = 3$ for $n \geq 2$. $P(n) = 3$ and $S(n) = 3n$.

We bring a finitely generated R-module A into the picture and define $H_{I,A}(n)$, the Hilbert function for I on A, by

$$H_{I,A}(n) = \lambda(I^nA/I^{n+1}A),$$

for any nonnegative integer n, where I is an $\underline{m}$-primary ideal in the local ring $(R,\underline{m})$. Recall that dimension A = dimension $(R/\text{ann } A)$.

2.1'. Hilbert Samuel Theorem. $H_{I,A}(n)$ is, for large n, a polynomial $P_{I,A}(n)$ in n with coefficients in $\mathbb{Q}$. The degree of $P_{I,A}(n)$ is $\delta-1$, where δ is the dimension of A. $\sum_{i=0}^{n} H_{I,A}(i)$ is, for large n, a polynomial $S_{I,A}(n)$ of degree δ. We write

$$S_{I,A}(n) = c_d\binom{n+d}{d} + c_{d-1}\binom{n+d-1}{d-1} + \cdots + c_0,$$

where $c_d,\ldots,c_0$ are integers. The multiplicity of I on A, $e(I;A)$, is defined by $e(I;A) = c_d$. Note that $e(I;A) = 0$ if and only if $\dim A < \dim R$. If $A = R/J$, for some ideal J in R, then $e(I;A) = 0$ unless $\dim A = \dim R$, in which case, $e(I;A) = e(IA)$.

3. Superficial elements.

When the local ring $(R,\underline{m})$ is Cohen-Macaulay, CM for short, there is a relationship between numbers of generators of certain ideals and $e(R)$. This is one reason for seeking methods for computing $e(R)$. The notion of superficial element is very useful in this regard.

3.1. Definition. Let $(R,\underline{m})$ be a local ring. Let

$$G_{\underline{m}}(R) = G(R) = R/\underline{m} \oplus \underline{m}/\underline{m}^2 \oplus \cdots$$

be the associated graded ring. An element x in $\underline{m}$ is superficial if there is a positive integer c such that

$$(\underline{m}^n : x) \cap \underline{m}^c = \underline{m}^{n-1},$$

for all $n > c$.

Thus x is a superficial element if and only if there is a positive integer c such that $\overline{x}$, the image of x in $\underline{m}/\underline{m}^2$ is non-zero and

$$(0 : \overline{x}G(R)) \cap \underline{m}^n/\underline{m}^{n+1} = 0,$$

for all $n \geq c$.

Good references for facts about superficial elements are Nagata's book [51] and Zariski and Samuel, volume II [70]. The following proposition is a slight variation of the usual existence theorem for superficial elements, cf. [51; (22.1)] or [70; VIII §8].

3.2. Proposition. Let $(R,\underline{m})$ be a local ring with $R/\underline{m}$ infinite. Let $I, J_1,\ldots,J_s$ be distinct ideals of R which are also distinct from $\underline{m}$. Then there is an element x in R such that

(i) $x \notin J_i$, $i = 1,\ldots,s$;

(ii) x is a superficial element for R

(iii) the image of x in R/I is a superficial element for R/I.

Proof. Let $G = R/\underline{m} \oplus \underline{m}/\underline{m}^2 \oplus \cdots$ and $\overline{G} = R/\underline{m} \oplus \underline{m}/\underline{m}^2 + I \oplus \underline{m}^2 + I/\underline{m}^3 + I \oplus \cdots$. Then $\overline{G} = G/K$ where K is a homogeneous ideal of G. Let Ass $G = \{P_1,\ldots,P_t\}$ and Ass $\overline{G} = \{Q_1/K,\ldots,Q_\ell/K\}$, where the Q_i are primes in G. Suppose that P_t is the maximal homogeneous ideal of G and that $Q_\ell = P_t/K$. The following subspaces are all proper $R/\underline{m}$ - subspaces of $\underline{m}/\underline{m}^2$:

$$J_i + \underline{m}^2/\underline{m}^2,\ i = 1,\ldots,s;\ P_i \cap \underline{m}/\underline{m}^2,\ i = 1,\ldots,t-1$$

$$Q_i \cap \underline{m}/\underline{m}^2,\ i = 1,\ldots,\ell - 1 \text{ and } I + \underline{m}^2/\underline{m}^2.$$

Since $R/\underline{m}$ is infinite, there is a nonzero $\overline{x}$ in $\underline{m}/\underline{m}^2$ such that $\overline{x}$ is not in any of these subspaces. We claim that if x is any element of $\underline{m}$ which maps to $\overline{x}$, then x is superficial for R and the image of x in R/I is superficial for R/I. We need to show that $(0 : \overline{x}G) \cap \underline{m}^n/\underline{m}^{n+1} = 0$ and $(0 : (x + I/I)\overline{G} \cap \underline{m}^n + I/\underline{m}^{n+1} + I = 0$ for all large n. Let $0 = N_1 \cap \cdots \cap N_{t-1} \cap N_t$, with N_i a P_i-primary ideal, be an irredundant primary decomposition of 0 in G. Then $(0 : xG) \subseteq N_1 \cap \cdots \cap N_{t-1}$. But $(\underline{m}/\underline{m}^2)^c \subseteq N_t$ for some nonnegative integer c, so that $(0 : \overline{x}G) \cap \underline{m}^n/\underline{m}^{n+1} = 0$ for $n \geq c$. The same reasoning shows that the image of x in R/I is a superficial element for R/I.

Remarks. (1) The Artin-Rees Lemma implies that there is an integer j such that $(\underline{m}^n : xR) \subseteq \underline{m}^{n-j} + (0 : xR)$ for all integers $n \geq j$. Thus if x is a superficial element and a nonzero divisor of R, $(\underline{m}^n : xR) = \underline{m}^{n-1}$, for all large n.

(2) If x is any element of $\underline{m}$, then

$$S_{\underline{m}/xR}(n) = S_{\underline{m}}(n) - \lambda(\underline{m}^n : xR).$$

If x is a superficial element for R, then

$$S_{\underline{m}/xR}(n) = S_{\underline{m}}(n) - S_{\underline{m}}(n-1) + \lambda(\underline{m}^c + (\underline{m}^n : xR)/\underline{m}^c),$$

where c is as in (3.1). Thus $S_{\underline{m}/xR}(n)$ differs from $S_{\underline{m}}(n) - S_{\underline{m}}(n-1)$ only in its constant term. Thus if $d = \dim R = 1$, $\lambda(R/xR) = e(R/xR) = e(R) + \lambda(0 : xR)$. If $d > 1$, $e(R/xR) = e(R)$.
Proposition 3.3 below is a strengthened version of [70; VIII, Thm. 22] but the proof is essentially the same.

3.3. Proposition. Let $(R,\underline{m})$ be a d-dimensional local ring with $R/\underline{m}$ an infinite field. Let I be an ideal of dimension r. Then there is a system of parameters $x_1,\ldots,x_d$ of R with the property that $e(R) = e((x_1,\ldots,x_d))$ and $e(R/I) = e((\overline{x}_1,\ldots,\overline{x}_r))$, where $\overline{x}_i$ denotes the image of x_i in R/I.

Proof. The proof is by induction on d. For the case $d = 0$, $e(R) = e((0))$. If $d = 1$, pick x as in (3.2) where the J_i are taken to be the isolated primes of 0 in R. We show that $e(R) = e((x))$ and, if $r = 1$ also, that $e(R/I) = e((x))$. If $\underline{m}$ does not belong to 0 in R then $(\underline{m}^n : xR) = \underline{m}^{n-1}$ for large n, so that $S_{\underline{m}/xR}(n) = S_{\underline{m}}(n) - S_{\underline{m}}(n-1) = e(R)$. For large n, $S_{\underline{m}/xR}(n) = \lambda(R/xR) = \lambda(x^nR/x^{n+1}R) = e((x))$.

Still in the case $d = 1$, if x is a zero divisor, let $J = (0 : xR)$. Then J has finite length. Pass to the ring $(R/J,\underline{m}/J)$. Then

$S_{\underline{m}/J}(n) = \lambda(R/\underline{m}^{n+1} + J) = \lambda(R/\underline{m}^{n+1}) - \lambda(\underline{m}^{n+1} + J/\underline{m}^{n+1}) =$ $\lambda(R/\underline{m}^{n+1}) - \lambda(J/\underline{m}^{n+1} \cap J) = \lambda(R/\underline{m}^{n+1}) - \lambda(J)$, for n large, because J has finite length, and $\bigcap_{t=0}^{\infty}(\underline{m}^t \cap J) = 0$. Thus $e(R) = e(R/J)$, and $e(x) = e(x + J/I)$, follows in a similar way since the polynomials involved are all of degree 1. Since $x + J/J$ is not a zero divisor in R/J, by the first part of the proof, $e(R/J) = e((x + J/J))$. The same argument may be applied to $\overline{x}$ to show that $e(R/I) = e((\overline{x}))$ if $r = 1$.

Now assume that $d > 1$. Pick x as in (3.2) where the J_i are the minimal primes of R and the minimal primes over I, in case $r \neq d$. Let $R^* = R/xR$ and $\underline{m}^* = \underline{m}/xR$. The polynomials $S_{\underline{m}^*}(n)$ and $S_{\underline{m}}(n) - S_{\underline{m}}(n-1)$ are of degree $d - 1$ and have the same leading term, so $e(R) = e(R^*)$. If $r = 1$, we use the argument above to show that $e(R/I) = e((x))$. We may assume that $r > 1$ and then we have that $e(R/I) = e(R^*/I^*)$, where $I^* = (I,x)/(x)$. By the induction hypothesis, there is a s.o.p. $x_2^*,\ldots,x_d^*$ for R^* such that $e(R^*) = e((x_2^*,\ldots,x_d^*))$, and $e(R^*/I^*) = e((\overline{x}_2^*,\ldots,\overline{x}_d^*))$, where $\overline{x}_i^*$ is the image of x_i^* in R^*/I^*. Pick x_i in R mapping onto x_i^* in R^*; then x, $x_2,\ldots,x_d$ is a s.o.p. for R. $e(R) = e((x_2^*,\ldots,x_d^*)) =$ $e((x,x_2,\ldots,x_d)/xR) \geq e((x,x_2,\ldots,x_d)) \geq e(R)$; similarly for $e(R/I)$.

For reference, we quote two results, more or less directly, from [70, VIII, §10].

3.4. Theorem. Let $(R,\underline{m})$ be a local ring and I the ideal generated by a s.o.p. $x_1,\ldots,x_d$. Then $e(I) \leq \lambda(R/I)$, and $e(I) = \lambda(R/I)$ if and only if $x_1,\ldots,x_d$ is an R-sequence, i.e., if and only if $G_I(R)$ is isomorphic to the polynomial ring $R/I[X_1,\ldots,X_d]$.

3.5. Theorem. Let $(A,\underline{p})$ be a local domain with quotient field K. Let $\underline{q}$ be a $\underline{p}$-primary ideal. Let $(R,\underline{m})$ be a local ring containing A with total quotient ring T, and let R be a finitely generated torsion free A-module. Then the polynomials

$[T : K]S_{\underline{q}}(n)$ and $[R/\underline{m} : A/\underline{p}]S_{\underline{q}R}(n)$ have the same degree and the same leading term.

Examples. (1). Let k be a field and let $A = k[[x^2,xy,y^2]]$. We may compute $e(A)$ as follows. Let $R = k[[x,y]]$. Then, with the notation as in (3.5), $[T : K] = 2$ and, by (3.5), $2e(A) = e((x^2,xy,y^2)R) = e(\underline{m}^2) = 4$. More generally, let $A = k[[(x_1,\ldots,x_d)^n]]$ for positive integers d and n. Let $R = k[[x_1,\ldots,x_d]]$. Then $[T : K] = n$ so $ne(A) = e(\underline{m}^n) = n^d$ and $e(A) = n^{d-1}$.

(2). If we find a s.o.p. $x_1,\ldots,x_d$ for the local ring $(R,\underline{m})$ such that $\underline{m}^n = (x_1,\ldots,x_d)\underline{m}^{n-1}$, for some n, then $e(R) = e((x_1,\ldots,x_d))$. In this case, the ideal $I = (x_1,\ldots,x_d)$ is called a minimal reduction of $\underline{m}$, cf. Northcott and Rees' paper [52] and Chapter 2 in these notes.

(3). Let k be a field and v an integer > 1. Let

$$R = k[[t^v,t^{v+1},\ldots,t^{2v-1}]].$$

Then $\underline{m} = (t^v,t^{v+1},\ldots,t^{2v-1})$, $\underline{m}^2 = t^v\underline{m}$, so $e(R) = e((t^v))$. But, by (3.4), $e(t^vR) = e(t^vk[[t]]) = v$.

(4). Let $R = k[[x^2,x^3,xy,y]]$, then $\underline{m} = (x^2,x^3,xy,y)$ and $\underline{m}^2 = (x^2,y)\underline{m}$. Thus $e(R) = e((x^2,y)) = e((x^2,y)k[[x,y]]) = 2$. Note that (x^2,y) is not an R-sequence so that $2 = e((x^2,y)) < \lambda(R/(x^2,y)) = 3$.

Thus (3.3) means that the computation of $e(R)$ can be reduced to the computation of $e((x_1,\ldots,x_d))$ for a system of parameters $x_1,\ldots,x_d$, for we may always assume that $R/\underline{m}$ is infinite by making the flat change of rings $R \to R(u) = R[u]_{\underline{m}R[u]}$, where u is an indeterminate. If I is an ideal of R, $v(I) = v(IR(u))$. If I is $\underline{m}$-primary, $\lambda(R/I) = \lambda(R(u)/IR(u))$ and $e(I) = e(IR(u))$. We will need the homological characterization of $e((x_1,\ldots,x_d))$ which arises from the Koszul complex associated to $x_1,\ldots,x_d$.

4. The Koszul complex, depth and multiplicity.

Let $\underline{x} = x_1,\ldots,x_n$ be a sequence of elements in the Noetherian ring R and let A be a finitely generated R-module. We recall that the Koszul complex $K(\underline{x})$ is the exterior algebra complex associated to R^n and the homomorphism $f\colon R^n \to R$ defined by $f(r_1,\ldots,r_n) = \sum_{i=1}^{n} r_i x_i$. We have $K_p(\underline{x}) \cong \wedge^p R^n$ with differential $d\colon K_p(\underline{x}) \to K_{p-1}(\underline{x})$ given by

$$d(e_1 \wedge \cdots \wedge e_p) = \sum_{i=1}^{p} (-1)^{i+1} x_i e_1 \wedge \cdots \wedge \hat{e}_i \wedge \cdots \wedge e_p,$$

where $e_1,\ldots,e_n$ is a basis for R^n. $K(\underline{x};\ A)$ is the complex $K(\underline{x}) \otimes A$ with differential $d \otimes 1$. For details see Serre's book [61]. We quote with proof the following theorem from Buchsbaum's paper [13].

4.1. Theorem. Let R be a Noetherian ring, A a finitely generated R-module and I an ideal of R generated by $x_1,\ldots,x_n$ such that $A/IA \neq 0$. Let $y_1,\ldots,y_s$ be a maximal A-sequence in I. Then

$$s + q = n,$$

where q is the dimension of the highest non-vanishing homology of the complex $K(\underline{x};\ A)$. Furthermore,

$$H_q(K(\underline{x};\ A)) \cong (y_1,\ldots,y_s)A \underset{A}{:} I/(y_1,\ldots,y_s)A.$$

Proof. The proof is by induction on s. If $s = 0$, then I consists of zero divisors on A so there is a nonzero element $a \in A$ such that $Ia = 0$. Thus $a \in H_n(K(\underline{x};\ A)) = (0 \underset{A}{:} I)$, so $q = n$. Assume $s > 0$ and consider the exact sequence:

$$0 \to A \xrightarrow{y_1} A \to \bar{A} \to 0,$$

where $\bar{A} = A/y_1A$. Taking homology, we get the exact sequence:

$$0 \to H_{\bar{q}}(K(\underline{x}; A)) \xrightarrow{y_1} H_{\bar{q}}(K(\underline{x}; A)) \to H_{\bar{q}}(K(\underline{x}; \bar{A})) \to \cdots\cdots\cdots,$$

where $\bar{q}$ is the dimension of the highest non-vanishing homology module of $K(\underline{x}; \bar{A})$. Now $y_1 \in I$ and $I \subseteq \operatorname{ann} H_i(K(\underline{x}; A))$ for all i, so $H_{\bar{q}}(K(\underline{x}; A)) = 0$ and $H_q(K(\underline{x}; A)) \cong H_{q-1}(K(\underline{x}; A))$. Thus $q = \bar{q} - 1$. Since $y_2, \ldots, y_s$ is a maximal A-sequence in I, by induction we have $s - 1 + \bar{q} = n$. Thus $s + q = n$. Finally, the last statement of the theorem follows because we have proved that $H_{\bar{q}}(K(\underline{x}; \bar{A})) \cong H_q(K(\underline{x}; A))$ and by induction we have that $H_{\bar{q}}(K(\underline{x}; \bar{A})) \cong (y_2, \ldots, y_s)\bar{A} \underset{\bar{A}}{:} I/(y_2, \ldots, y_s)\bar{A}$

$$\cong (y_1, y_2, \ldots, y_s)A \underset{A}{:} I/(y_1, y_2, \ldots, y_s)A.$$

4.2. Corollary. Let R be a local ring. Let I be an ideal containing an R-sequence of length s and let $\underline{x} = x_1, \ldots, x_n$ be a minimal basis for I. Then $H_p(K(\underline{x})) = 0$ for $p > n - s$, and $H_{n-s}(K(\underline{x})) \cong \operatorname{Ext}^s(R/I, R)$.

Proof. Let $y_1, \ldots, y_s$ be a maximal R-sequence in I, then by (4.1), $H_{n-s}(K(\underline{x})) \cong (y_1, \ldots, y_s) : I/(y_1, \ldots, y_s) \cong \operatorname{Hom}_R(R/I, R/(y_1, \ldots, y_s)) \cong \operatorname{Ext}^s(R/I, R)$.

Next, the Euler-Poincare characteristic of the Koszul complex is defined. It expresses the connection between Koszul complexes and multiplicities.

4.3. Definition. Let $(R, \underline{m})$ be a local ring, A a finitely generated R-module and I an $\underline{m}$-primary ideal. Then $(I, \operatorname{ann} A) \subseteq \operatorname{ann} H_i(K(\underline{x}; A))$, for all i, so that $H_i(K(\underline{x}; A))$ has finite length for all i. $\chi(\underline{x}; A)$, the Euler-Poincaré characteristic of $\underline{x}$ on A, is defined by

$$\chi(\underline{x}; A) = \sum_{i=1}^{\infty} (-1)^i \lambda(H_i(K(\underline{x}; A))).$$

4.4. Theorem. Let $(R, \underline{m})$ be a local ring and $\underline{x} = x_1, \ldots, x_d$ a system of parameters for R. Let A be a finitely generated R-module. Then

$$\chi(\underline{x};\ A) = e(\underline{x}R;\ A).$$

The proof is given in Serre's book [61; IV, Thm. 1] also, with more detail in Auslander and Buchsbaum's paper [5].

The following two well-known results are often used in the sequel.

4.5. Proposition. Let $(R,\underline{m})$ be a local ring and

$$0 \to A \to B \to C \to 0$$

an exact sequence of finitely generated R-modules.

(i) If depth B $<$ depth C, then depth A = depth B.

(ii) If depth B $>$ depth C, then depth A = depth C + 1.

(iii) If depth B = depth C, then depth A $\geq$ depth B.

This is [39; ex. 14, §3.1]; a different proof can be obtained using the characterization of depth via the functor Ext.

4.6. Theorem. Let $(R,\underline{m})$ be a local ring and A a finitely generated R-module of finite projective dimension. Then

$$\text{depth } R = \text{depth } A + \text{pd } A.$$

(4.6) is due to Auslander and Buchsbaum [6] and Serre [62]; a proof can also be found in [39] and [51].

We will need to use the fact that an ideal of finite projective dimension in a local ring contains a nonzero divisor. This follows from the following theorem of Auslander and Buchsbaum [5]. A proof can be found in [39].

4.7. Theorem. Let $(R,\underline{m})$ be a local ring. Let A be a finitely generated R-module of finite projective dimension.

Then either ann A = 0 or ann A contains a nonzero divisor.

Counting generators is easier in CM rings. The result, found in EGA [30], that, under "geometric conditions", the CM locus is Zariski-open will be useful. We have taken (4.8) from Hochster's monograph [35].

4.8. Lemma. Let M be a finitely generated module over a Noetherian ring R. Let V_i or $V_i(M)$ denote the set

$$\{P \in \mathrm{Spec}(R) \mid \mathrm{pd}_{R_P} M_P \geq i\}.$$

Then V_i is Zariski-closed. If we let

$$J_i(M) = \{r \in R \mid \mathrm{pd}_{R_r} M_r < i\}$$

then $J_i(M)$ is a radical ideal and V_i is the set of primes containing $J_i(M)$.

Note. The convention is that the zero module has projective dimension -1.

Proof of (4.8). In case $i = 0$, V_0 is the support of M and $J_0(M) = \mathrm{ann}\, M$. If $i = 1$, the assertion that V_1 is closed is equivalent to the statement that the set of primes $P \in \mathrm{Spec}(R)$ such that M_P is R_P-free is open. Suppose that M_P is R_P-free. Then there is a map

$$\varphi: R^t \to M$$

for some t such that φ becomes an isomorphism upon applying $\otimes_R R_P$. Then there is an $r \in R \setminus P$ which kills both $\ker \varphi$ and $\mathrm{coker}\, \varphi$. Thus $M \otimes_R R_r$ is free and M_Q is free for Q not containing r.

If $i > 1$, consider an exact sequence

$$0 \to M' \to F \to M \to 0$$

where F is a finitely generated free R-module. For $i > 1$,

$V_i(M) = V_{i-1}(M')$, and the result that $V_i(M)$ is closed follows by induction.

Now, $r \in \bigcap_{P \in V_i} P$ if and only if, for all Q not containing r, $pd_{R_Q} M_Q < i$. But

$$pd_{R_r} M_r = \sup\{pd_{R_Q} M_Q \mid Q \text{ does not contain } r\}.$$

So $r \in \bigcap_{P \in V_i} P$ if and only if $pd_{R_r} M_r < i$ if and only if $r \in J_i(M)$.

4.9. Proposition. Let R be a homomorphic image of a regular Noetherian domain. Then the Cohen-Macaulay locus of R is Zariski-open.

Proof. Write $R = S/I$, where S is a regular Noetherian domain, and I is an ideal of S. Let $P \in \mathrm{Spec}(R)$. $P = Q/I$, with $Q \in \mathrm{Spec}(S)$. $\dim R_P - \mathrm{depth}\, R_P = \dim S_Q - \mathrm{height}\, I_Q - \dim S_Q + \mathrm{pd}\, S_Q$ (by (4.6)). Thus

$$\dim R_P - \mathrm{depth}\, R_P = \mathrm{pd}\, S_Q - \mathrm{height}\, I_Q.$$

Height I_Q takes at most a finite number of values for $Q \in \mathrm{Spec}(S)$, so it follows from (4.8) that the set

$$\{P \in \mathrm{Spec}(R) \mid \dim R_P - \mathrm{depth}\, R_P = 0\}$$

is Zariski-open. But this set is the Cohen-Macaulay locus of R.

5. Cohen-Macaulay and Gorenstein quotients of local Gorenstein rings.

Theorem 5.2 below is due to Serre and is found in Bass' Ubiquity paper [7].

5.1. Proposition. Let $(R,\underline{m})$ be a d-dimensional local Gorenstein ring and I an ideal of height h. R/I is Cohen-Macaulay if and only if $\mathrm{Ext}^h_R(R/I,R) \neq 0$ and $\mathrm{Ext}^i_R(R/I,R) = 0$

for $i \neq h$.

Proof. Since I contains an R-sequence of length h, $\mathrm{Ext}_R^i(R/I,R) = 0$ for $i < h$. The fact that $\mathrm{Ext}_R^h(R/I,R)$ is the highest nonvanishing $\mathrm{Ext}_R^i(R/I,R)$ exactly when R/I is Cohen-Macaulay follows from [39; Thms. 217, 218].

5.2. Theorem. Let $(R,\underline{m})$ be a d-dimensional local Gorenstein ring and I an ideal of height h. Then R/I is Gorenstein if and only if R/I is Cohen-Macaulay and $\mathrm{Ext}_R^h(R/I, R) \cong R/I$.

Proof. Let $\Omega = \mathrm{Ext}_R^h(R/I, R)$. First we reduce to the case where $h = 0$. If $h \neq 0$, let $z_1,\ldots,z_h = \underline{z}$ be a maximal R-sequence in I. Let $\overline{R} = R/\underline{z}R$. Then $\overline{R}$ is a local Gorenstein ring of dimension $d - h$, $\overline{I} = I/\underline{z}R$ is an ideal of height 0 and $\Omega = \mathrm{Ext}_R^h(R/I, R) \cong \mathrm{Hom}_{R/\underline{z}R}(R/I,R/\underline{z}R)$. Thus we may assume that $h = 0$. If we denote by $E_R(A)$, the R-injective envelope of the R-module A, then, cf. [39, Th. 203], $E_{R/I}(R/\underline{m}) = \mathrm{Hom}_R(R/I, E_R(R/\underline{m}))$. If $d = 0$, $E_R(R/\underline{m}) = R$, so $\Omega = E_{R/I}(R/\underline{m})$, and $E_{R/I}(R/\underline{m}) \cong R/I$ if and only if R/I is Gorenstein. Assume that $d > 0$. Let $\underline{x} = x_1,\ldots,x_d$ be a maximal R-sequence. Then, since R/I is CM of dimension d, $\underline{x}$ is also a maximal R/I-sequence. It follows easily by induction that $\Omega/\underline{x}\Omega = \mathrm{Hom}_{R/\underline{x}R}(R/(I,\underline{x}), R/\underline{x}R)$. Thus, by the case $d = 0$, we have $R/(I,\underline{x})$ is Gorenstein if and only if $R/(I,\underline{x})$ is CM and $\Omega/\underline{x}\Omega \cong R/(I,\underline{x})$. But R/I is Gorenstein if and only if $R/(I,\underline{x})$ is Gorenstein, and, by Nakayama's lemma, $\Omega/\underline{x}\Omega$ is cyclic if and only if Ω is.

With the notation as above, $\Omega = \Omega_{R/I}$ is called the canonical module for R/I, see Herzog and Kunz [33]. Combining (5.2) with (4.2) we have the following result of Kunz [41].

5.3. Corollary. Let R be a d-dimensional local Gorenstein ring and I an ideal of height h, with I minimally generated by $x_1,\ldots,x_n$. Then $H_{n-h}(K(\underline{x})) \cong \mathrm{Ext}_R^h(R/I,R) = \Omega_{R/I}$.

6. Exactness of complexes of finitely generated free modules.

Let R be a commutative ring and $f\colon F \to G$ a homomorphism of finitely generated free R-modules F and G. Recall that rank $f = \sup\{r \mid \wedge^r f \neq 0\}$. If we fix bases for F and G and let $[f]$ be the corresponding matrix, then rank f is the size of the largest submatrix of $[f]$ with nonvanishing determinant. We denote by $\mathscr{I}(f)$ the ideal generated by all minors of order n where n = rank f. In this section we recall McCoy's theorem [49] which shows how to test for injectivity of f and provides a crucial nonzero divisor when f is a monomorphism. Then we turn to results about complexes of finitely generated free modules: namely, the acyclicity lemma of Peskine and Szpiro [53] and the Buchsbaum and Eisenbud characterization of when such a complex is exact [17].

6.1. <u>McCoy's Theorem</u>. Let R be a commutative ring and $f\colon F \to G$ a homomorphism of finitely generated free R-modules. f is a monomorphism if and only if rank f = rank F and $(0 : \mathscr{I}(f)) = 0$.

A proof can be found in Kaplansky's book [39]; see also the paper [25] of Eagon and Northcott where the spirit is somewhat like the original [49].

6.2. <u>Lemma</u>. Let $f\colon F \to G$ be a homomorphism of finitely generated free modules F and G over a commutative ring R. Then coker f is projective if and only if $\mathscr{I}(f)$ is generated by an idempotent.

<u>Proof</u>. Since $\mathscr{I}(f)$ is finitely generated, $\mathscr{I}(f)$ is generated by an idempotent if and only if $\mathscr{I}(f)$ is locally trivial, i.e., locally 0 or 1. Hence, also using the fact that coker f is finitely presented, we may localize at a prime ideal P where rank f = rank f_P. So we assume that R is quasi-local. If coker f is projective, it is free and splits off. So $F \cong F_1 \oplus \ker f$, $G \cong F_1 \oplus \operatorname{coker} f$ and f is the map $1_{F_1} \oplus 0$. Thus $\mathscr{I}(f) = R$. Conversely, suppose that rank $f = r$ and $\mathscr{I}(f) = R$. Since R is quasi-local, some

$r \times r$ minor of f is a unit. We may choose bases for F and G so that the matrix of f has the form

$$[f] = \begin{bmatrix} 1_r & B \\ C & D \end{bmatrix}$$

where 1_r is the $r \times r$ identity matrix. By elementary row and column operations we can make $B = 0$ and $C = 0$; but then $D = 0$ since rank $f = r$. Clearly, coker f is free.

6.3. Lemma. Let R be a commutative ring. Let

$$\mathbb{F}\colon F \xrightarrow{f} G \xrightarrow{g} H$$

be a complex of finitely generated free R-modules with $\mathscr{I}(f) = \mathscr{I}(g) = R$. $\mathbb{F}$ is exact if and only if rank $G =$ rank f + rank g.

Proof. We may assume that R is quasi-local. By (6.2) we may choose bases for F, G and H so that

$$[f] = \begin{bmatrix} 1_r & 0 \\ 0 & 0 \end{bmatrix}, \text{ and } [g] = \begin{bmatrix} 0 & 0 \\ 0 & 1_s \end{bmatrix},$$

where rank $f = r$ and rank $g = s$. It is now clear that $\mathbb{F}$ is exact if and only if rank $G = r + s$.

(6.4) is the "acyclicity lemma" of Peskine and Szpiro [53].

6.4. Lemma. Let $(R,\underline{m})$ be a local ring of depth r. Let

$$\mathbb{F}\colon 0 \to F_n \to F_{n-1} \to \cdots \to F_0$$

be a complex of finitely generated free R-modules, where $n \leq r$. Assume that the homology of $\mathbb{F}$, $H_i(\mathbb{F})$, has depth 0 for $i > 0$. Then $H_i(\mathbb{F}) = 0$ for $i > 0$.

Proof. We may assume $r > 0$. $0 \to H_n(\mathbb{F}) \to F_n$ is exact, so $H_n(\mathbb{F}) = 0$ since depth $F_n = r > 0$. Thus we may assume that $r > 1$. Let j be an integer such that $H_i(\mathbb{F}) = 0$ for $i > j$. If $j = 0$, there is nothing to prove. We assume that $j > 0$ and show $H_j(\mathbb{F}) = 0$. Consider the following exact sequences:

$$\text{(i)} \quad 0 \to F_n \to F_{n-1} \to \cdots \to F_{j+1} \to C \to 0$$

$$\text{(ii)} \quad 0 \to Z \to F_j \to F_{j-1}$$

$$\text{(iii)} \quad 0 \to C \to Z \to H_j(F) \to 0.$$

By (4.6) and (i), depth $C \geq 2$. By (ii), depth $Z \geq 1$, so (4.5) applied to (iii) gives depth $C = 1$ unless $H_j(F) = 0$.

The following theorem of Buchsbaum and Eisenbud [17] is just one of their fundamental structure theorems on finite free resolutions.

6.5. Theorem. Let R be a Noetherian ring. Let

$$\mathbb{F}: 0 \to F_n \xrightarrow{f_n} F_{n-1} \to \cdots \to F_1 \xrightarrow{f_1} F_0$$

be a complex of finitely generated free R-modules. F is exact if and only if, for all j, $1 \leq j \leq n$,

(a) rank f_{j+1} + rank f_j = rank F_j

(b) $\mathscr{I}(f_j)$ contains an R-sequence of length j or $\mathscr{I}(f_j) = R$.

Proof. We show first that (a) and (b) imply that $\mathbb{F}$ is exact. The idea of the proof is to use (6.2) and (6.3) to reduce to the situation in (6.4). We may assume that R is local; say depth $R = r$. Then (b) implies that $\mathscr{I}(f_n) = \cdots = \mathscr{I}(f_{r+1}) = R$. Let $C = \operatorname{coker} f_{r+1}$. By (6.2), C is free and, by (6.3), the complex

$$\mathbb{F}: \quad 0 \to F_n \to F_{n-1} \to \cdots \to F_{r+1} \to F_r \to C \to 0$$

is exact. It remains to prove that the complex

$$\mathbb{F}': \quad 0 \to C \to F_{r-1} \to F_{r-2} \to \cdots \to F_1 \to F_0$$

is exact. Let $d = \dim R$. If $d = 0$, $r = 0$ and there is nothing to prove. Assume $d > 0$. Then, by induction on d, we have for, $i > 0$, that $H_i(\mathbb{F}') \otimes R_P = 0$ for all non-maximal primes P. Thus depth $H_i(\mathbb{F}') = 0$ for $i > 0$, and we apply (6.4) to show that $\mathbb{F}'$ is exact.

For the converse, assume that $\mathbb{F}$ is exact. We show that (a) holds. By (6.1), $\mathscr{I}(f_n)$ contains a nonzero divisor. Let S be the multiplicatively closed set it generates. Then $\mathbb{F} \otimes R_S$ is exact and may be spliced into the exact complexes:

$$0 \to F_n \otimes R_S \to F_{n-1} \otimes R_S \to C \to 0,$$

where $C = \operatorname{coker} f_n \otimes 1$ is free, and

$$0 \to C \to F_{n-2} \otimes R_S \to F_{n-3} \otimes R_S \to \cdots \to F_0 \otimes R_S.$$

(a) then follows for $\mathbb{F}_S$, and thus for $\mathbb{F}$, by (6.3) for the first complex and by induction on the length of the complex for the second.

We must now prove that (b) holds. We do this by induction on n, the length of the complex. If $n = 1$, (6.1) gives the required nonzero divisor in $\mathscr{I}(f_1)$. Assume that $n > 1$. Suppose that $\mathscr{I}(f_n)$ satisfies (b) but that some $\mathscr{I}(f_j)$, for $j < n$ fails (b). Let $x_1, \ldots, x_\ell$ be a maximal R-sequence in $\mathscr{I}(f_j)$, $0 \leq \ell < j < n$, and let $P_1, \ldots, P_t$ be the prime ideals belonging to $(x_1, \ldots, x_\ell)R$. $\mathscr{I}(f_n) \nsubseteq \bigcup_{i=1}^{t} P_i$ so there is a nonzero divisor y in $\mathscr{I}(f_n) \setminus \bigcup_{i=1}^{t} P_i$. If we localize at the multiplicatively closed set S generated by y, we reduce the length of the complex by (6.2) and preserve the hypothesis on $\mathscr{I}(f_j)$, i.e., $x_1, \ldots, x_\ell$ is a maximal R_S-sequence in $\mathscr{I}(f_j \otimes R_S) = \mathscr{I}(f_j) \otimes R_S$. So now we assume that $\mathscr{I}(f_n)$ fails (b). Then, we may (cf. [39], Thm. 135) assume that R is local of depth $\ell < n$. The following lemma then finishes the proof.

6.6. Lemma. Let $(R,\underline{m})$ be a local ring of depth ℓ. Let

$$\mathbb{F}: \quad 0 \to F_n \xrightarrow{f_n} F_{n-1} \to \cdots \to F_1 \xrightarrow{f_1} F_0$$

be an exact complex of finitely generated free R-modules. If $n > \ell$, then the map f_n splits, i.e., $\mathscr{I}(f_n) = R$.

Proof. Suppose $\mathscr{I}(f_n) \subseteq \underline{m}$. Then, by (6.1), $\ell \geq 1$. Let C_j = image f_j = ker f_{j-1}, for $j = 2,\ldots,n$. Then $C_n = F_n$ and, since C_{n-1} is not free, (4.5) implies that depth $C_{n-1} = \ell - 1$. We have the exact sequences

$$0 \to C_{n-j+1} \to F_{n-j} \to C_{n-j} \to 0,$$

for $j = 1,\ldots,n-1$. It follows from (4.5) that depth $C_{n-j} = \ell - j$. But then $C_{n-\ell}$ has depth 0. This is impossible since $0 \to C_{n-\ell} \to F_{n-\ell-1}$ is exact.

We will also need the following information from Buchsbaum and Eisenbud's paper [16].

6.7. Proposition. With the notation as in (6.5) assume that the given complex is exact. Then, for all $j \geq 1$,

$$\sqrt{\mathscr{I}(f_j)} \subseteq \sqrt{\mathscr{I}(f_{j+1})}.$$

Proof. Let s be a nonzero divisor in $\sqrt{\mathscr{I}(f_j)}$ and let S be the multiplicatively closed set generated by s. Then, $\mathscr{I}((f_j)_S) = (\mathscr{I}(f_j))_S = R_S$, so by (6.2), coker $(f_j)_S$ is projective over R_S. It follows that coker $(f_{j+1})_S$ is projective over R_S and $\mathscr{I}(f_{j+1})_S = (\mathscr{I}(f_{j+1}))_S = R_S$. Thus $s \in \sqrt{\mathscr{I}(f_{j+1})}$.

7. Two computations of heights.

We will need a result of Eagon [24] giving an upper bound for heights of the ideals $\mathscr{I}_t(A)$, the ideals generated by the minors of order t of a matrix A over a Noetherian ring R. Much more is now known about these ideals. The reader is referred to [24], [36], [38], [43].

7.1. Theorem. Let R be a Noetherian ring and $A = (a_{ij})$ and $r \times s$, $s \leq r$, matrix with entries in R. Let $t \leq s$. If $\mathscr{I}_t(A)$ is a proper ideal,

$$\text{height } \mathscr{I}_t(A) \leq (r - t + 1)(s - t + 1).$$

Proof. The proof is by induction on s. If $t = 1$, $\mathscr{I}_t(A)$ is generated by rs elements so height $\mathscr{I}_t(A) \leq rs$. Thus we may assume $t > 1$ and $s > 1$. We also may assume R is local with maximal ideal $\underline{m}$ and that $\mathscr{I}_t(A)$ is $\underline{m}$-primary. The conclusion follows immediately by induction unless all the a_{ij} are in $\underline{m}$, so we assume this to be the case. Let u be an indeterminate and pass to the ring $R[u]$. Let B be the matrix

$$B = \begin{bmatrix} a_{11} + u & a_{12} & \cdots & a_{1s} \\ a_{21} & a_{22} & & a_{2s} \\ \vdots & \vdots & & \vdots \\ a_{r1} & a_{r2} & & a_{rs} \end{bmatrix}$$

$\mathscr{I}_t(B) \subseteq \underline{m}R[u]$, since $t > 1$. We claim that $\underline{m}R[u]$ is minimal over $\mathscr{I}_t(B)$. For if $Q \subseteq \underline{m}R[u]$ is a prime ideal minimal over $\mathscr{I}_t(B)$, then

$$(\mathscr{I}_t(B), u) = (\mathscr{I}_t(A)R[u], u) \subseteq (Q, u) \subseteq (\underline{m}R[u], u).$$

Thus $(\underline{m}R[u], u)$ is minimal over (Q, u) so that $(\underline{m}R[u], u)/Q$ has height 1 in $R[u]/Q$. It follows that $Q = \underline{m}R[u]$. Now we can transfer the problem to the matrix B and the ring $R(u) = R[u]_{\underline{m}R[u]}$. The proof follows by induction because $a_{11} + u$ is a unit in $R(u)$.

We will also need the theorem of Serre on the dimension of the intersection of affine varieties. The proof is found in [61; III, Prop. 17].

7.2. Theorem. Let k be a field. Let I and J be

two ideals in the polynomial ring $k[X_1,\ldots,X_n]$. Then

$$\text{height } (I + J) \leq \text{height } (I) + \text{height } (J).$$

Serre has proved that the same result holds for ideals in any regular local ring. The proof is much harder and is found in [61; V, Thm. 3].

8. Coherence, uniform coherence and descent of flatness.

It may seem odd that we need results about coherent rings in these notes. It turns out, however, that one way to handle questions about numbers of generators of intersections of ideals is to change them into questions about numbers of generators of relation modules for maps $R^n \to R$, with n fixed, and in order to bound these we need to pass from the ring R to the ring $\prod_{\alpha \in X} R_\alpha = R^X$, for some set X.

We recall the definition of coherent module.

8.1. Definition. Let R be a ring and A a finitely generated R-module. A is coherent if every finitely generated submodule of A is finitely related. The ring R is coherent if R is a coherent R-module, i.e., if every finitely generated ideal of R is finitely related.

The following characterization of coherence comes from [20] and is found in Bourbaki [11] Ch. I, §2, exercise 12.

8.2. Proposition. Let R be a ring. R is coherent if and only if any direct product of copies of R is flat over R.

8.3. Definition. A ring R is uniformly coherent, if, for any positive integer n and any nonzero homomorphism $f: R^n \to R$, ker f can be generated by $\varphi(n)$ elements, where $\varphi(n)$ is a nonnegative integer depending only on n.

Remark. The notion of uniform coherence is due to Soublin [64] as is (8.4).

8.4. Proposition. A ring R is uniformly coherent if and only if $R^{\mathbb{N}}$ is coherent, where $\mathbb{N}$ is the set of natural numbers.

Proof. If R is uniformly coherent and if $f: (R^{\mathbb{N}})^{\ell} \to R^{\mathbb{N}}$ is any nonzero homomorphism, then $f = (f_i)_{i \in \mathbb{N}}$, where $f_i: R^{\ell} \to R$. Since $\ker f = \Pi \ker f_i$, we see that $\ker f$ is finitely generated as a $R^{\mathbb{N}}$-module.

Suppose that R is not uniformly coherent. Then there is an integer n and, for each $i \in \mathbb{N}$, a nonzero homomorphism $f_i: R^n \to R$ such that $\ker f_i$ cannot be generated by i elements. Let $f = (f_i)_{i \in \mathbb{N}}: (R^{\mathbb{N}})^n \to R^{\mathbb{N}}$. Then $\ker f$ is not finitely generated and R is not coherent.

8.5. Proposition. Let $\rho: R \to S$ be a homomorphism of rings making S an R-module of finite presentation. If R is coherent, so is S. If R is uniformly coherent, so is S.

Proof. By (8.4) it is sufficient to prove that R coherent implies that S is coherent. Clearly, it is sufficient to show that S is a coherent R-module. But this follows from the exact sequence

$$0 \to C \to R^n \to S \to 0,$$

since C is a finitely generated submodule of the coherent R-module R^n.

We need, under suitable conditions, to be able to descend the property of uniform coherence. The theorem we need, (8.6) below, is due to Quentel [54].

8.6. Theorem. Let $\rho: R \to S$ be a homomorphism of rings making S an R-module of finite presentation. Assume that $\ker \rho$ is a nil ideal of finite presentation. Then S coherent implies that R is coherent and S uniformly coherent implies that R is uniformly coherent.

Some machinery is required for the proof of (8.6) because

we are going to use (8.2) to change (8.6) into a question of descent of flatness. The proof of (8.6) can be reduced to two cases: (1) ρ is an epimorphism, (2) ρ is a monomorphism. For (1) we will need results on descent of flatness under an epimorphism with kernel a nilpotent ideal (8.8). For (2) we will need results on descent of flatness under a finite monomorphism (8.10).

8.7. Proposition. Let R be a ring and A an R-module of finite presentation. Then, for any family $(F_\alpha)_{\alpha \in X}$ of R-modules, the homomorphism

$$A \otimes \prod_{\alpha \in X} F_\alpha \to \prod_{\alpha \in X} A \otimes F_\alpha$$

is an isomorphism.

This is Bourbaki [11, Ch. 1, §2, ex. 9].

8.8. Theorem. Let R be a ring and I a nilpotent ideal. A module A is R-flat if and only if A/IA is R/I-flat and the canonical homomorphism $I \otimes A \to IA$ is an isomorphism.

This is part of the "local criterion of flatness" due to Grothendieck. A proof can be found in [11, Ch. 3, §5, Thm. 1].

For descent of flatness under a finite monomorphism we change the question of flatness into one of injectivity and use the following result of Eisenbud [27].

8.9. Theorem. Let $R \subset S$ be rings with S a finitely generated R-module. Let Q be any R-module. If $\text{Hom}_R(S,Q)$ is S-injective, then Q is R-injective.

Proof. Let E be the R-injective envelope of Q. The map u: $\text{Hom}_R(S,Q) \to \text{Hom}_R(S,E)$ induced by $Q \to E$ is a monomorphism of S-modules. Since $\text{Hom}_R(S,Q)$ is S-injective, u splits. We will show first that u is an epimorphism. To do this, it is sufficient to show that u is essential. Since S is finitely generated over R, we have an epimorphism $R^n \to S$ which induces the vertical monomorphisms in the following commutative diagram:

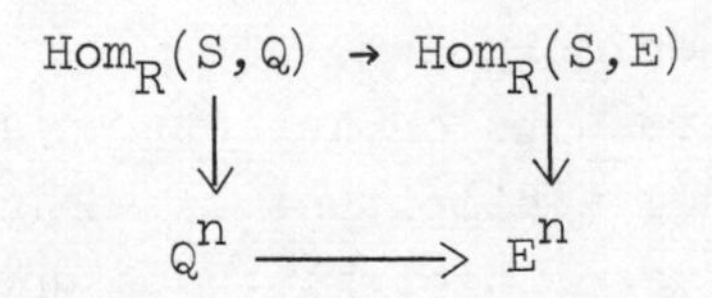

$Q \to E$ essential implies that $Q^n \to E^n$ is essential. Thus u is essential, since $\mathrm{Hom}_R(S,Q) = \mathrm{Hom}_R(S,E) \cap Q^n$. But we also have the commutative diagram:

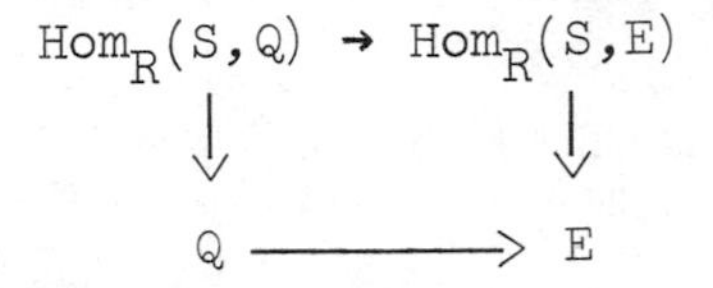

where the vertical maps are induced by the inclusion $R \to S$. The vertical map on the right is an epimorphism by the injectivity of E, hence $Q \to E$ is an epimorphism.

We now come to the theorem of Raynaud and Gruson [55] on descent of flatness under a finite monomorphism. The proof given below - in the same spirit as that in [55] - is due to Matlis.

8.10. <u>Theorem</u>. Let $R \subset S$ be rings with S a finitely generated R-module. Let A be an R-module. If $S \otimes_R A$ is S-flat, then A is R-flat.

Proof. Let E be the injective envelope of the direct sum of $R/\underline{m}_\alpha$ over all maximal ideals $\underline{m}_\alpha$ of R. Then, for any R-module C, $\mathrm{Hom}_R(C,E) = 0$ if and only if $C = 0$. Let T be any R-algebra and B any T-module. By [19, VI, Prop. 5.1].

$\mathrm{Ext}_T(D,\mathrm{Hom}_R(B,E)) \cong \mathrm{Hom}_R(\mathrm{Tor}^T(D,B),E)$, for all T-modules D. Thus it follows that B is T flat if and only if $\mathrm{Hom}_T(B,E)$ is T-injective. We apply this first with $T = S$ and then with $T = R$. Let $B = S \otimes_R A$. B is S-flat, so $\mathrm{Hom}_R(S \otimes_R A,E) \cong \mathrm{Hom}_R(S,\mathrm{Hom}_R(A,E))$ is S-injective. Thus, by

(8.9), $\mathrm{Hom}_R(A,E)$ is R-injective and A is R-flat.

Proof of (8.6). By (8.4), we need only prove that S coherent implies R coherent. We may reduce to the cases: (1) ρ is an epimorphism and (2) ρ is a monomorphism.

Assume that ρ is an epimorphism. We have the exact sequence

$$0 \to I \to R \to R/I \cong S \to 0,$$

where I is a nil ideal of finite presentation. Let X be any set, and let $A = R^X$. Since I and R/I are of finite presentation, by (8.7), $I \otimes_R R^X \cong IR^X$, and $R/I \otimes_R R^X \cong (R/I)^X$. Since R/I is coherent, $(R/I)^X$ is R/I-flat so the required flatness of R^X follows from (8.8).

Assume that ρ is a monomorphism. If X is any set, by (8.7), $S \otimes_R R^X \cong S^X$. S coherent implies that $S \otimes_R R^X$ is S-flat; then R^X is R-flat by (8.10).

CHAPTER 2
SOME COMBINATORIAL RESULTS

In this chapter we consider some general results about numbers of generators of ideals in arbitrary local rings which are combinatorial in nature. For consistency and ease of exposition, we generally keep our running Noetherian hypothesis, although the reader will note that with proper interpretation this is often unnecessary.

1. Extending bases.

Given the local ring $(R,\underline{m})$ and ideals $J \subset I$, a natural question to ask is: when can a minimal basis for J be extended to a minimal basis for I? (Note that pre-images of any minimal basis for I/J can always be extended to a minimal basis for I.)

1.1. Proposition. Let $(R,\underline{m})$ be a local ring and $J \subseteq I$ ideals of R. The following conditions are equivalent.

(1) Any minimal basis can be extended to a minimal basis for I.

(2) $\underline{m}I \cap J = \underline{m}J$.

(3) There is an ideal L of R such that $(I/J)/L(I/J)$ is free over R/L and $LI \cap J = LJ$.

(4) $v(I) = v(J) + v(I/J)$.

Proof. The equivalence follows immediately from the exact sequence

$$0 \to LI + J/LI \to I \otimes R/L \to I/J \otimes R/L \to 0,$$

and the isomorphism $LI + J/LI \cong J/LI \cap J$ which hold for any ideals $J \subseteq I$ and L in the ring R.

Chevalley's theorem characterizing regular quotients of regular local rings gives one instance of (1.1). Davis [23] generalized this as follows.

1.2. Theorem. Let $(R,\underline{m})$ be a local ring and $J \subseteq I$ ideals of R. If I/J is generated by a regular sequence, then

$$v(I) = v(J) + v(I/J).$$

Proof. Since I/J is generated by a regular sequence, $(I/J)/(I/J)^2 = (I/J)/I(I/J)$ is free over R/I. Hence we may apply (1.1) as soon as we check that $I^2 \cap J = IJ$. This follows from the following lemma.

1.3. Lemma. Let $(R,\underline{m})$ be a local ring and J an ideal of R. Let $x_1,\ldots,x_n$ be elements of R which form a regular sequence modulo J. Then, for all integers $t \geq 0$,

$$(x_1,\ldots,x_n)^t \cap J = (x_1,\ldots,x_n)^t J.$$

Proof. Let "-" denote reduction modulo J. Let t be any nonnegative integer. Let $f(x_1,\ldots,x_n) \in (x_1,\ldots,x_n)^t \cap J$, where $f(X_1,\ldots,X_n)$ is a form of degree t in the polynomial ring $R[X_1,\ldots,X_n]$. Then $f(x_1,\ldots,x_n) \in J$ implies that $\overline{f}(\overline{x}_1,\ldots,\overline{x}_n) = 0$. Since $(\overline{x}_1,\ldots,\overline{x}_n)^t/(\overline{x}_1,\ldots,\overline{x}_n)^{t+1}$ is free over $\overline{R}/(\overline{x}_1,\ldots,\overline{x}_n)$, all the coefficients of $\overline{f}$ must be in $(\overline{x}_1,\ldots,\overline{x}_n)$. Thus $f \in (x_1,\ldots,x_n)^t J + (x_1,\ldots,x_n)^{t+1} \cap J$. Consequently, $f \in (x_1,\ldots,x_n)^t J$.

The first corollary is due to Lichtenbaum and Schlessinger [44].

1.4. Corollary. If both I and I/J are generated by regular sequences then so is J.

1.5. Corollary. If R/J is regular then J is generated by part of a minimal basis for the maximal ideal of R. If R and R/J are regular, J is generated by a subset of a regular system of parameters for R.

A variation on this theme of extending bases arises in the work of Northcott and Rees [52] on reductions of ideals. Given an ideal I in a local ring $(R,\underline{m})$, they find inside I an ideal $\mathscr{X}$ having the same radical as I and having other properties in common with I, such that a minimal basis for $\mathscr{X}$ can be extended to a minimal basis for I.

1.6. Definition. If I and J are ideals in the local ring $(R,\underline{m})$, then J is a reduction of I if $J \subseteq I$ and $JI^n = I^{n+1}$ for some positive integer n. J is called a minimal reduction of I if J does not properly contain a reduction of I, this is equivalent to: J does not have any proper reductions.

Remark. If J is a reduction of I, then I and J have the same radical. If I is $\underline{m}$-primary and J is a reduction of I, then $e(J) = e(I)$.

Northcott and Rees [52] give the following proof of the existence of minimal reductions.

1.7. Theorem. Let $(R,\underline{m})$ be a local ring. Let J be a reduction of an ideal I of R. Then J contains a minimal reduction $\mathscr{X}$ of I. Moreover, if L is any ideal, $\mathscr{X} \subseteq L \subseteq I$, then any minimal basis of $\mathscr{X}$ can be extended to a minimal basis of L.

Proof. Let $\mathscr{S}$ be the set of all ideals $L + \underline{m}I$, where $L \subseteq J$ and L is a reduction of I. $\mathscr{S} \neq \emptyset$. Since $L + \underline{m}I/\underline{m}I$ is a subspace of $I/\underline{m}I$, $\mathscr{S}$ has a minimal element $U + \underline{m}I$. Let $x_1,\ldots,x_n$ be elements of U whose images modulo $\underline{m}I$ form a basis for $U + \underline{m}I/\underline{m}I$ over $R/\underline{m}$. Let $\mathscr{X} = (x_1,\ldots,x_n)$. From $\mathscr{X} + I\underline{m} = U + I\underline{m}$ it easily follows that $\mathscr{X}$ is also a reduction of I. We claim that $\mathscr{X}$ is a minimal reduction of I. By choice, $\mathscr{X} \cap I\underline{m} = \mathscr{X}\underline{m}$. Suppose $\mathscr{X}_0 \subseteq \mathscr{X}$ is a reduction of I. Then, by choice of U, $\mathscr{X}_0 + I\underline{m} = \mathscr{X} + I\underline{m}$. Let $x \in \mathscr{X}$, then $x = y + z$ with $y \in \mathscr{X}_0$ and $z \in I\underline{m}$. Thus $z \in I\underline{m} \cap \mathscr{X} = \mathscr{X}\underline{m}$, so $x \in \mathscr{X}_0 + \mathscr{X}\underline{m}$ and $\mathscr{X} \subseteq \mathscr{X}_0$.

(1.7) gives an indication of a relationship between numbers of generators and multiplicities - a recurring theme throughout these notes.

Example. It follows that if I has a reduction $\mathscr{X}$ which is generated by height I elements, then $\mathscr{X}$ is a minimal reduction of I. If, say, $R = k[[x^2,xy,y^2]]$, k a field, then $\underline{m}^2 = (x^2,y^2)\underline{m}$, so that (x^2,y^2) is a minimal reduction for $\underline{m}$. We will see in the example following (2.2) that the number of elements in a minimal reduction of an ideal I can be greater than height I.

In the next section we investigate the connection between reductions and numbers of generators of powers of an ideal.

2. Numbers of generators of powers of an ideal.

We seek information about numbers of generators of powers of an ideal I in a local ring $(R,\underline{m})$. The following simple fact gives some insight into one method of attack on the problem. Suppose J and K are subsets of R with $J \subseteq I$ and $K \subseteq I^{t-1}$ for some positive integer t. If $I^t = JKR$, then $I^{t+1} = J^2KR$. In particular, if J is a reduction of I, say $I^t = JI^{t-1}$, then $v(I^{t+s}) \leq v(I^{t-1})\cdot v(J^{s+1})$, for all integers $s \geq 0$.

We have a Hilbert function defined as follows:

$$H_{I,\underline{m}}(n) = v(I^n) = \dim_{R/\underline{m}}(I^n/I^n\underline{m}),$$

for all nonnegative integers n. For large n, $H_{I,\underline{m}}(n)$ is a polynomial $P_{I,\underline{m}}(n)$ with rational coefficients (cf. Chapter 1). Let $\deg(P_{I,\underline{m}}(n)) = \ell(I) - 1$.

Northcott and Rees [52] have shown, cf. (2.2) below, in the case that $R/\underline{m}$ is infinite, that if $\mathscr{X}$ is a minimal reduction for I and $x_1,\ldots,x_r$ is a minimal basis for $\mathscr{X}$, then, not only can $x_1,\ldots,x_r$ be extended to a minimal basis of I, but also, for any positive integer n, the monomials of degree n in $x_1,\ldots,x_r$ can be extended to a minimal basis of

I^n. In particular, this gives a lower bound for the number of generators of I^n. This property of $x_1,\ldots,x_r$ is made precise in Definition (2.1) below and then the theorem of Northcott and Rees follows.

2.1. Definition. Let $x_1,\ldots,x_r$ be elements of an ideal I in the local ring $(R,\underline{m})$. $x_1,\ldots,x_r$ are analytically independent in I if whenever $f(X_1,\ldots,X_r)$ is a form of (arbitrary) degree t in the polynomial ring $R[X_1,\ldots,X_r]$ such that $f(x_1,\ldots,x_r) \equiv 0 \bmod I^t\underline{m}$, then all the coefficients of f are in $\underline{m}$.

Thus, if $x_1,\ldots,x_r$ are analytically independent in I and and $\mathscr{X} = (x_1,\ldots,x_r)$, then

$$H_{\mathscr{X},\underline{m}}(n) = P_{\mathscr{X},\underline{m}}(n) = \binom{n+r-1}{r-1} \leq H_{I,\underline{m}}(n).$$

2.2. Theorem. Let $(R,\underline{m})$ be a local ring with infinite residue field. Let I be an ideal of R and $\mathscr{X}$ a reduction of I with minimal basis $x_1,\ldots,x_r$. Then $\mathscr{X}$ is a minimal reduction of I if and only if $x_1,\ldots,x_r$ are analytically independent in I and $r = \ell(I)$.

Proof. Assume that $x_1,\ldots,x_r$ is a minimal basis of a minimal reduction $\mathscr{X}$ of I. Let $f(X_1,\ldots,X_r)$ be a form of degree t in the polynomial ring $R[X_1,\ldots,X_r]$ and suppose that $f(x_1,\ldots,x_t) \equiv 0 \bmod I^t\underline{m}$. Suppose that the coefficient of X_1^t is not in $\underline{m}$. Then $x_1^t \in (x_2,\ldots,x_r)^{t-1} + I^t\underline{m}$ so that $\mathscr{X}^t \subseteq (x_2,\ldots,x_r)\mathscr{X}^{t-1} + I^t\underline{m}$. Since $\mathscr{X}$ is a reduction of I, there is an integer s such that $\mathscr{X}I^s = I^{s+1}$. Hence, $I^{s+t} = I^s\mathscr{X}^t \subseteq (x_2,\ldots,x_r)I^{s+t-1} + I^{s+t}\underline{m}$, and $I^{s+t} = (x_2,\ldots,x_r)I^{s+t-1}$. But this contradicts the fact that $x_1,x_2,\ldots,x_r$ is a minimal basis of $\mathscr{X}$ and $\mathscr{X}$ is a minimal reduction of I. Thus the coefficient of X_1^t is in $\underline{m}$.

Now we show that if $a_{11},a_{21},\ldots,a_{r1}$ are any elements of R not all in $\underline{m}$, then $f(a_{11},a_{21},\ldots,a_{r1}) \equiv 0 \bmod \underline{m}$. Thus, since $R/\underline{m}$ is an infinite field, all the coefficients of f must be in $\underline{m}$. We change the basis of $\mathscr{X}$ as follows. Pick

$\underline{\text{Proof}}$. We may assume that $R/\underline{m}$ is infinite. If $\dim R = 0$, I is nilpotent and $\ell(I) = 0$. Assume $\dim R > 0$. By (2.2) we may assume that I is generated by an analytically independent set of elements $x_1,\ldots,x_r$ with $r = \ell(I)$.

We first reduce to the case where R is a domain. Let $P_1,\ldots,P_t$ be the minimal prime ideals in R. Since, $\lambda(I^n/\underline{m}\, I^n) \geq \lambda(I^n/\underline{m}\, I^n + P_i \cap I^n)$, we have that $\ell(I) \geq \ell(I + P_i/P_i)$. If $\ell(I) > \ell(I + P_i/P_i)$ for $i = 1,\ldots,t$, then, for each i, there is a homogeneous form $F_i(X_1,\ldots,X_r)$ in the variables $X_1,\ldots,X_r$ with some coefficient not in $\underline{m}$ such that $F_i(x_1,\ldots,x_r) \in P_i$. Hence

$$(F_1(x_1,\ldots,x_r) \cdots F_t(x_1,\ldots,x_r))^N = 0,$$

for some N. But this is impossible since $x_1,\ldots,x_r$ are analytically independent and some coefficient of $(F_1\cdots F_t)^N$ is not in $\underline{m}$.

We assume that R is a domain with quotient field K. Let $R^I = R[I/x_1R]$. Then $R^I/\underline{m}R^I$ is isomorphic to a polynomial ring in $r - 1$ variables over $R/\underline{m}$ by the analytic independence of $x_1,\ldots,x_r$. Now, from [70, p. 326] we have

$$\text{height } \underline{m}R^I + r-1 \leq \text{height } \underline{m}.$$

Thus $r \leq \dim R$.

Thus we have a lower bound for the number of generators of powers of an ideal I in a local ring $(R,\underline{m})$ in terms of $\ell(I)$:

$$v(I^n) \geq \binom{n + \ell(I) - 1}{\ell(I) - 1}.$$

A different approach to the problem of estimating $v(I^n)$ is taken by Eakin and Sathaye in [26]. They prove an existence theorem for reductions of an ideal I based on rather broad estimates for the number of generators of I^n, thereby getting estimates for $v(I^{n+s})$. Eakin and Sathaye's proof of additional elements a_{ij}, $1 \leq i \leq r$, $2 \leq j \leq r$, so that the determinant $|a_{ij}|$ is a unit in R and define a new minimal basis $y_1,\ldots,y_r$ for $\mathscr{X}$ by

$$x_i = \sum_{j=1}^{r} a_{ij}y_j$$

Then, $f(\sum_{j=1}^{r} a_{1j}y_j,\ldots,\sum_{j=1}^{r} a_{rj}y_j) \equiv 0 \bmod I^t\underline{m}$. By what we have already shown, $f(\sum_{j=1}^{r} a_{1j}X_j,\ldots,\sum_{j=1}^{r} a_{rj}X_j)$ has coefficient of X_1^t in $\underline{m}$, i.e., $f(a_{11},\ldots,a_{r1}) \equiv 0 \bmod \underline{m}$. This completes the proof that $x_1,\ldots,x_r$ are analytically independent in I. It follows that $H_{I,\underline{m}}(n) \geq \binom{n+r-1}{r-1}$ and thus that $\ell(I) - 1 \geq r - 1$. Since $\mathscr{X}$ is a reduction of I, $\mathscr{X}^s I^t = I^{t+s}$, for some positive integer t and all integers $s \geq 0$. Thus $v(I^{t+s}) \leq v(I^t)\binom{s+r-1}{r-1}$, for all such s, and $\ell(I) - 1 \leq r - 1$. Hence $\ell(I) = r$.

For the converse, assume that $\mathscr{Y} \subseteq \mathscr{X}$ is a minimal reduction of I. Then $\mathscr{Y}$ has a minimal basis of $\ell(I)$ analytically independent elements, by what we have just shown, and this minimal basis can, by (1.7), be extended to a minimal basis of $\mathscr{X}$. Hence $\mathscr{Y} = \mathscr{X}$.

Example. $\ell(\underline{m}) = \dim R$, and in general we have

$$\text{height } I \leq \ell(I) \leq \dim R.$$

The latter inequality will be proved below. It may happen that height $I < \ell(I)$. Let $R = k[[x^2,xy,y^2]]$, k a field. Let $I = (x^2,xy)$. x^2,xy are analytically independent in I, so that I itself is a minimal reduction of I and $\ell(I) = 2 >$ height I.

The following proof is taken from [18] where it is attributed to Rees.

2.3 Theorem. Let $(R,\underline{m})$ be a local ring and I an ideal. Then

$$\ell(I) \leq \dim R.$$

this result ((2.3) below) requires passage to a non-Noetherian ring for "elbow room," so for the remainder of §2 we work in a quasi-local ring. The definition of $v(I)$ for an ideal I is the same as before, only now $v(I)$ may be infinite.

2.3. Theorem. Let $(R,\underline{m})$ be a quasi-local ring with infinite residue field. Let I be an ideal of R. Let n and r be positive integers. If

$$v(I^n) < \binom{n+r}{r},$$

then there are elements $y_1,\ldots,y_r$ in I such that $I^n = (y_1,\ldots,y_r)I^{n-1}$.

We need three lemmas for the proof of (2.3). Note that by Nakayama's lemma, for information about $v(I^n)$ we may pass to the ring $R/\underline{m}I^n$, so that - changing notation - I^n itself is a vector space over $R/\underline{m}$.

2.4. Lemma. Let $(R,\underline{m})$ be a quasi-local ring and I an ideal such that $\underline{m}I^n = 0$. Let $y \in I$ and let $K = (0 : yR) \cap I^{n-1}$. Denote the map $R \to R/K$ by "-". If T is an ideal $\subseteq I^{n-1}$, then $\overline{T} \cong yT$.

Proof. $\overline{T} = T + K/K \cong T/T \cap K \cong yT$.

2.5. Lemma. Let $(R,\underline{m})$ be a quasi-local ring and I an ideal such that for some positive integers n and r, $\underline{m}I^n = 0$ and $v(I^n) = \dim_{R/\underline{m}} I^n < \binom{n+r}{r}$. Suppose that there is a sequence of elements $\{y_i\}_{i=1}^{\infty}$ in I and an action of S_ω, the group of permutations of the natural numbers, as a group of automorphisms of R in such a way that for each σ in S_ω, $\sigma(I) = I$ and $\sigma(y_i) = y_{\sigma(i)}$. If $I^n = \sum_{i=1}^{r+1} y_i I^{n-1}$, then $I^n = \sum_{i=1}^{r} y_i I^{n-1}$.

(Note that the introduction of the S_ω action means that any sequence of t y's is "as good as" any other.)

Proof. Suppose the lemma is false for some positive integers r and n. Pick r minimal and n minimal for this r. Let δ be the common $R/\underline{m}$-vector space dimension of all $y_i I^{n-1}$. There are two possibilities.

Case 1. $\delta \geq \binom{n+r-1}{r}$. For this case, $r = 1$ is impossible. For then $y_i I^{n-1}$ would have dimension at least n. But I^n has dimension at most n, so we would have the contradiction $I^n = y_i I^{n-1}$. Thus we have $r > 1$. Since $\binom{n+r-1}{r} + \binom{n+r-1}{r-1} = \binom{n+r}{r}$, by counting lengths, we see that in the ring $\overline{R} = R/y_1 I^{n-1}$, the ideal $\overline{I}^n$ has at most $\binom{n+r-1}{r-1}$ generators. It is straightforward to check that $\overline{R}$ satisfies the hypotheses of the lemma with $\{\overline{y}_i\}_{i=2}^{\infty}$ taking the place of $\{y_i\}_{i=1}^{\infty}$ and S_ω acting on $\overline{R}$ via the induced actions from all $\sigma \in S_\omega$ such that $\sigma(y_1) = y_1$. Then, by the minimality of r, $\overline{I}^n = \sum_{i=2}^{r} \overline{y}_i \overline{I}^{n-1}$. In other words, $I^n = \sum_{i=2}^{r} y_i I^{n-1} + y_1 I^{n-1} = \sum_{i=1}^{r} y_i I^{n-1}$, a contradiction. Thus we must have

Case 2. $\delta < \binom{n+r-1}{r}$. Let $K = (0 : y_1) \cap I^{n-1}$ and let "-" denote the map $R \to R/K$. Then, by (2.4), $\overline{I}^{n-1} \cong y_1 I^{n-1}$. $\overline{R}$ also satisfies the hypotheses of the lemma when n is changed to $n - 1$. For $\underline{m}\overline{I}^{n-1} \cong \underline{m}y_1 I^{n-1} = 0$, and $\dim_{R/\underline{m}} \overline{I}^{n-1} = \dim_{R/\underline{m}} y_1 I^n < \binom{n-1+r}{r}$, by the hypothesis of Case 2. The set $\{\overline{y}_i\}_{i=2}^{\infty}$ is a sequence of elements in $\overline{I}$ and $\{\sigma \in S_\omega \mid \sigma(y_1) = y_1\}$ is a copy of S_ω which acts on $\overline{R}$ in the prescribed manner since such a σ maps K onto itself. By the minimality of n,

$$\overline{I}^{n-1} = \sum_{i=2}^{r+1} \overline{y}_i \overline{I}^{n-2}.$$

Thus $$I^{n-1} = \sum_{i=2}^{r+1} y_i I^{n-2} + K,$$

so $$y_1 I^{n-1} = \sum_{i=2}^{r+1} y_1 y_i I^{n-2}.$$

But $I^n = \sum_{i=1}^{r+1} y_i I^{n-1}$. Thus $I^n = \sum_{i=2}^{r+1} y_i I^{n-1}$. By hypothesis, there is no preference for any r of the y's, so $I^n = \sum_{i=1}^{r} y_i I^{n-1}$, a contradiction. This completes the proof of the lemma.

2.6. Lemma. Let $(R,\underline{m})$ be a quasi-local ring and I an ideal such that for some positive integers n and r, $\underline{m}I^n = 0$ and $v(I^n) = \dim_{R/\underline{m}} I^n < \binom{n+r}{r}$. Suppose that there is a sequence of elements $\{y_i\}_{i=1}^{\infty}$ in I and a positive integer s such that, for all i, $\{y_i,\ldots,y_{i+s-1}\}$ is a generating set for I. Assume also that S_ω acts as a group of automorphisms of R in such a way that for each $\sigma \in S_\omega$, $\sigma(y_i) = y_{\sigma(i)}$. Then $I^n = (y_i,\ldots,y_{i+r-1})I^{n-1}$, for each i.

Proof. The set $\mathscr{S}$ of all positive integers j such that

$$I^n = (y_1,\ldots,y_{i+j-1})I^{n-1}, \text{ for all } i$$

is non-empty by hypothesis. Let ℓ be the smallest element. (2.5) implies that $\ell \leq r$.

Proof of (2.3). Let $x_1,\ldots,x_t$ be a set of elements of I such that $I^n = (x_1,\ldots,x_t)^n$. We may in fact assume that $I = (x_1,\ldots,x_t)$. Let $\{u_{ij}\}$, for $1 \leq i < \infty$, $1 \leq j \leq t$, be independent indeterminates and pass to the ring $R(u) = R(\{u_{ij}\}) = R[\{u_{ij}\}]_{\underline{m}R[\{u_{ij}\}]}$. For $1 \leq i < \infty$, define the elements y_i in $IR(u)$ by

$$y_i = \sum_{j=1}^{t} u_{ij} x_j.$$

Let S_ω act on $R(u)$ by $\sigma(u_{ij}) = u_{\sigma(i)j}$. Thus $\sigma(y_i) = y_{\sigma(i)}$. We claim that, for each i,

$$I^n R(u) = \sum_{j=0}^{r-1} y_{i+j} I^{n-1} R(u)$$

By Nakayama's lemma we may prove this under the additional assumption that $\underline{m} I^n R(u) = 0$. Now we are in the situation described by (2.6) so the claim holds.

Let A_q be the monomials of degree n-1 in $x_1,\ldots,x_t$ and B_ℓ the monomials of degree n in $x_1,\ldots,x_t$. By the above claim (with $s = 1$), there is a set of equations

$$(*) \qquad B_\ell = \sum_{j=1}^{r} y_j \sum_q h_{jq} A_q, \text{ with } h_{jq} \in R(u).$$

By comparing coefficients we can assume that the h_{jq} involve only the u_{ij} which occur in the definition of $y_1,\ldots,y_r$. Call these rt variables U_{ij}. Let $f(U_{ij})$ be a common denominator for the h_{ij}. Since $R/\underline{m}$ is infinite, we can find a_{ij} in R, $1 \leq i \leq r$, $1 \leq j \leq t$, with $f(a_{ij}) \notin \underline{m}$. Then we can specialize the set of equations (*) to conclude that

$$I^n = \sum_{j=1}^{r} \left(\sum_{i=1}^{t} a_{ij} x_i \right) I^{n-1}.$$

Note. The proof shows that r "generic" linear combinations of $x_1,\ldots,x_t$ will serve for the y's.

Example. Let $(R,\underline{m})$ be a one dimensional local CM ring. It follows from (2.3), since $v(\underline{m}^j) \leq e(R)$ for all $j \geq 0$, cf. (Chapter 3; 1.1) that $v(\underline{m}^{e(R)-1}) = H(e(R)-1) = e(R)$. More generally, it follows from (2.3) that if $v(\underline{m}^t) = H(t) \leq t$, then there is a y in $\underline{m}$ such that $\underline{m}^t = y\underline{m}^{t-1}$. This means that $H(n) = H(t-1)$ for all $n \geq t - 1$. Thus we must have $v(\underline{m}^{t-1}) = H(t-1) = e(R)$. For example, let k be a field and take $R = k[[x^5,x^8,x^{27}]]$. We have $H(1) = H(2) = 3$. $H(3)$, which must be greater than 3 since $e(R) = 5$, is in fact equal to 4. Then $H(n) = H(4) = 5$, for all $n \geq 4$.

3. The Hilbert function for $\underline{m}$.

Relatively little is known about the behavior of the Hilbert function $H(n)$ for the maximal ideal in a local ring $(R,\underline{m})$ - much less for more general ideals I. We would like to find answers to such questions as: "When is $H(n)$ a nondecreasing function?", "When does $H(n)$ become a polynomial?" etc.

Consider, for example, the case of a one dimensional local CM ring $(R,\underline{m})$. It is well known, cf. (Chapter 3; 1.1) that $v(\underline{m}^n) = H(n) \leq e(R)$, for all $n \geq 0$. From (2.3) it follows that $H(n) = e(R)$ for all $n \geq e(R) - 1$. Only recently have examples been given showing that $H(n)$ can decrease. Herzog and Waldi [34] give the following example.

<u>Example</u>. Let k be a field. Let

$$R = k[[t^{30},t^{35},t^{42},t^{47},t^{148},t^{153},t^{157},t^{169},t^{181},t^{193}]].$$

Then $\underline{m}^2 = (t^{60},t^{65},t^{70},t^{72},t^{77},t^{82},t^{84},t^{89},t^{94})$, so $H(1) = 10$ and $H(2) = 9$.

Eakin and Sathaye [26] give the example

<u>Example</u>. Let k be a field. Let

$$R = k[[t^{15},t^{21},t^{23},t^{47},t^{48},t^{49},t^{50},t^{52},t^{54},t^{55},t^{56},t^{58}]].$$

Then $H(1) = 12$ and $H(2) = 11$.

However, it seems reasonable that for one dimensional local CM rings of small enough (say, at most three?) embedding dimension, $H(n)$ is nondecreasing. Matlis [47] proved that if $(R,\underline{m})$ is a one dimensional local CM ring of embedding dimension two, then $H(n)$ is nondecreasing, in fact, $H(n) = n + 1$, for $n < e(R)$ and $H(n) = e(R)$ for $n \geq e(R) - 1$. Boratynski and Swiecicka [10] have a generalization which is given in (3.6) below. Note that the example $k[[t^4,t^5,t^{11}]]$ shows that $H(n)$ need not increase for $n < e(R) - 1$, even for embedding dimension three.

Let $(R,\underline{m})$ be a local ring. If x is a superficial element for R, then $e(R) = e(R/xR)$ provided that x is a non-zero divisor or $\dim R > 1$. We would like to compare Hilbert functions for R and R/xR, not only for x superficial but also for any x in $\underline{m}$. The natural way to do this is to compare H_R^0 and $H_{R/xR}^1$. (For notation and definitions, cf. (Chapter 1; 2)).

Examples. Let k be a field. Let $R = k[[t^2,t^3]]$. t^2 is a superficial element for R. We have

$$H_R^0(0) = 1 = H_{\overline{R}}^1(0)$$

$$H_R^0(n) = 2 = H_{\overline{R}}^1(n), \text{ for all } n > 0,$$

where $\overline{R} = R/t^2R$.

Let $S = k[[t^4,t^5,t^{11}]]$. t^4 is a superficial element for S. We have

$$H_S^0(0) = 1 \qquad H_{\overline{S}}^1(0) = 1$$

$$H_S^0(1) = 3 \qquad H_{\overline{S}}^1(1) = 3$$

$$H_S^0(2) = 3 \qquad H_{\overline{S}}^1(2) = 4$$

$$H_S^0(n) = 4 \qquad H_{\overline{S}}^1(n) = 4, \text{ for all } n \geq 3,$$

where $\overline{S} = S/t^4S$.

The reason for the discrepancy between H_S^0 and $H_{\overline{S}}^1$ is that the image of t^4 in $\underline{m}/\underline{m}^2$ is a zero divisor in $G(S)$, the associated graded ring.

The obstruction to the equality of H_R^0 and $H_{R/xR}^1$ is named in the following theorem.

3.1. Theorem. Let $(R,\underline{m})$ be a local ring and let $x \in \underline{m}$. For every $n \geq 0$,

$$H^1_{R/xR}(n) - H^0_R(n) = \lambda((\underline{m}^{n+1}: xR)/\underline{m}^n).$$

Proof. First note that $H^1_R(n) - H^1_{R/xR}(n) =$ $\lambda(R/(\underline{m}^{n+1}: xR))$. For $\lambda(R/\underline{m}^{n+1}) - \lambda(R/\underline{m}^{n+1} + Rx) =$ $\lambda(\underline{m}^{n+1} + Rx/\underline{m}^{n+1}) = \lambda(Rx/\underline{m}^{n+1} \cap Rx)$. Now the homomorphism $R \to xR/(m^{n+1} \cap Rx) \to 0$ induced by multiplication by x has kernel $(\underline{m}^{n+1}: xR)$. Hence $\lambda(Rx/\underline{m}^{n+1} \cap Rx) = \lambda(R/(\underline{m}^{n+1}: xR))$. Thus, $H^1_{R/xR}(n) - H^0_R(n) = H^1_{R/xR}(n) - H^1_R(n) + H^1_R(n-1) =$ $- \lambda(R/(\underline{m}^{n+1}: xR)) + \lambda(R/\underline{m}^n) = \lambda((\underline{m}^{n+1}: xR)/\underline{m}^n)$.

Remark. The result (3.1) is found in Singh's paper [63]. (3.1) is the special case for local rings of a more general result which appears in Stanley's paper [65] and which Stanley attributes to Macaulay.

Note. It follows immediately from (3.1) that if $x \in \underline{m} \setminus \underline{m}^2$, then $H^0_R(n) = H^1_{R/xR}(n)$, for all $n \geq 0$, if and only if the image of x in $G(R)$, the associated graded ring, is a nonzero divisor. We extend this remark in Corollary (3.3) but we prove the following lemma first.

3.2. Lemma. Let $(R,\underline{m})$ be a local ring and let $x \in \underline{m} \setminus \underline{m}^2$. Let $\overline{x}$ be the image of x in $G(R)$. If $\overline{x}$ is a nonzero divisor in $G(R)$, then $G(R/xR) \cong G(R)/\overline{x}G(R)$.

Proof. x is a nonzero divisor in $G(R)$ if and only if $(\underline{m}^n: xR) = \underline{m}^{n-1}$ for all $n > 0$, and this is true if and only if $\underline{m}^n \cap xR = x\underline{m}^{n-1}$. Now,

$$G(R/xR)_n = \underline{m}^n + xR/\underline{m}^{n+1} + xR \cong \underline{m}^n/\underline{m}^n \cap \underline{m}^{n+1} + xR$$

$$\cong \underline{m}^n/\underline{m}^{n+1} + \underline{m}^n \cap xR,$$

and $(G(R)/xG(R))_n = \underline{m}^n/x\underline{m}^{n-1} + \underline{m}^{n+1}$, so the required conclusion follows.

3.3. Corollary. Let $(R,\underline{m})$ be a d dimensional local ring. Let $x_1,\ldots,x_t$ be elements of $\underline{m}\setminus\underline{m}^2$ with images $\overline{x}_1,\ldots,\overline{x}_t$ in $G(R)$ forming a $G(R)$-sequence. Then,

$$H_R^0 = H^t_{R/(x_1,\ldots,x_t)R}.$$

In particular, if $G(R)$ is Cohen-Macaulay, then there is an Artin local ring R^* of embedding dimension $v(\underline{m}) - d$ and length $e(R)$, such that

$$H_R^0 = H^d_{R^*}.$$

Proof. The image of x_i in $G(R/(x_1,\ldots,x_{i-1})R) \cong G(R)/(\overline{x}_1,\ldots,\overline{x}_{i-1})G(R)$ is a nonzero divisor if and only if $(\underline{m}^{n+1},x_1,\ldots,x_{i-1})\colon x_iR = (\underline{m}^n, x_1,\ldots,x_{i-1})$, for all $n \geq 0$. Since $x_1,\ldots,x_t$ is a $G(R)$-sequence, (3.1) gives that,

$$H^0_{R/(x_1,\ldots,x_{i-1})R}(n) = H^1_{R/(x_1,\ldots,x_i)R}(n)$$

for all $n \geq 0$, and all i, $1 \leq i \leq t$. Thus,

$$H_R^0 = H^t_{R/(x_1,\ldots,x_t)R}.$$

Suppose that $G(R)$ is Cohen-Macaulay. Then, by passing to $R(u)$, where u is an indeterminate, if necessary, there is a maximal $G(R)$-sequence $x_1,\ldots,x_d$ of elements of degree 1. Let $x_1,\ldots,x_d$ be pre-images in R, and let $R^* = R/(x_1,\ldots,x_d)R$. R^* is an Artin local ring. Since $x_1,\ldots,x_d$ is a reduction of $\underline{m}$, R^* has embedding dimension $v(\underline{m}) - d$ and length $e(R)$, cf. (1.7). By the first part of the corollary we have

$$H_R^0 = H^d_{R^*}.$$

3.4. Proposition. Let $(R,\underline{m})$ be a d dimensional local Cohen-Macaulay ring of embedding dimension at most $d + 1$. Then $G(R)$ is Cohen-Macaulay.

Proof. We may assume that R is complete and that

$v(\underline{m}) = d + 1$. By the Cohen structure theory, $R = S/fS$, where S is a regular local ring of dimension $d + 1$ and f is in the square of the maximal ideal of S. By (3.2), $G(R) \cong G(S)/\overline{f}G(S)$, where $\overline{f}$ is the image of f in $G(S)$. Thus $G(R)$ is CM.

(3.5) is due to Boratynski and Swiecicka [10] and (3.6) to Matlis [47].

3.5. Theorem. Let $(R,\underline{m})$ be a d dimensional local Cohen-Macaulay ring of embedding dimension at most $d + 1$. Then there is an Artin local ring R^* of embedding dimension at most 1 and length $e(R)$, such that

$$H_R^0 = H_{R^*}^d.$$

Proof. This follows from (3.3) and (3.4).

3.6. Corollary. Let $(R,\underline{m})$ be a 1 dimensional local Cohen-Macaulay ring of embedding dimension 2. Then,

$$H_R^0(n) = n + 1, \text{ for } n \leq e(R) - 1,$$

$$H_R^0(n) = e(R), \quad \text{for } n \geq e(R).$$

Proof. Let R^* be the Artin local ring from (3.5). R^* has principal maximal ideal and $\lambda(R^*) = e(R)$. Thus

$$H_R^0(n) = H_{R^*}^1(n) = \sum_{i=0}^{n} H_{R^*}^0(i) = \begin{cases} n + 1, & \text{if } n < \lambda(R^*) \\ \lambda(R^*) & \text{if } n \geq \lambda(R^*). \end{cases}$$

Remark. (3.6) also follows directly from (3.4) and the fact that $G(R)$ CM implies that $H_R^0(n)$ is strictly increasing if $H_R^0(n) < e(R)$.

Another result like (3.5) is

3.7. Theorem. Let $(R,\underline{m})$ be a local Cohen-Macaulay ring of dimension d. If R has embedding dimension $e(R) + d - 1$, then there is an Artin local ring R^* of embedding dimension

$e(R) - 1$ and length $e(R)$ such $H_R^0 = H_{R*}^d$. In fact,

$$H_R^0(n) = e\binom{n + d - 2}{n - 1} + \binom{n + d - 2}{m}.$$

<u>Proof</u>. This will follow from (3.3) once we show, as we do in (3.10) below, that $G(R)$ is Cohen-Macaulay.

First we see that local Cohen-Macaulay rings of embedding dimension $e(R) + d - 1$ are characterized by the fact that if the associated graded ring $G(R)$ has a homogeneous system of parameters of degree 1, $\overline{x}_1,\ldots,\overline{x}_d$, then $\mathcal{M}^2 \subseteq (\overline{x}_1,\ldots,\overline{x}_d)G(R)$, where $\mathcal{M}$ is the maximal homogeneous ideal of $G(R)$.

3.8. <u>Theorem</u>. Let $(R,\underline{m})$ be a d dimensional local Cohen-Macaulay ring with infinite residue field. Then there exist elements $x_1,\ldots,x_d$ in $\underline{m}$ such that $\underline{m}^2 = (x_1,\ldots,x_d)\underline{m}$ if and only if $v(\underline{m}) = e(R) + d - 1$.

<u>Proof</u>. By (2.2), there exist elements $x_1,\ldots,x_d$ in $\underline{m}$ with $\underline{m}^n = \underline{xm}^{n-1}$ for some integer n, where $\underline{x} = (x_1,\ldots,x_d)$. Tensor the exact sequence

$$0 \to \underline{m} \to R \to R/\underline{m} \to 0$$

by $R/\underline{x}$ to obtain the exact sequence

$$0 \to \mathrm{Tor}_1^R(R/\underline{m}, R/\underline{x}) \to \underline{m}/\underline{xm} \to R/\underline{x} \to R/\underline{m} \to 0.$$

We have that $\lambda(\mathrm{Tor}_1^R(R/\underline{m}, R/\underline{x})) = d$ because $\underline{x}$ is a regular sequence in the Cohen-Macaulay ring R. By the properties of minimal reductions, $e(R) = \lambda(R/\underline{x})$. Hence $\lambda(\underline{m}/\underline{xm}) = e(R) + d - 1$. From this it follows immediately that $v(\underline{m}) \leq e(R) + d - 1$ and that $v(\underline{m}) + d - 1$ if and only if $\underline{m}^2 = \underline{xm}$.

(3.9) below is adapted for the special case under our consideration from a theorem and proof of Hochster and Ratliff [37].

3.9. <u>Theorem</u>. Let $(R,\underline{m})$ be a local ring and $G(R)$ its associated graded ring. $G(R)$ is Cohen-Macaulay if and only if $G(R)_{\mathcal{M}}$ is Cohen-Macaulay, where $\mathcal{M}$ is the maximal homogeneous ideal of $G(R)$.

<u>Proof</u>. The condition is clearly necessary. To prove sufficiency we may assume that $k = R/\underline{m}$ is an infinite field. We have that $G(R) \cong k[X_1,\ldots,X_s]/\mathcal{J}$, where $\mathcal{J}$ is a homogeneous ideal in the polynomial ring $k[X_1,\ldots,X_s]$. By (Chapter 1; 4.9], the Cohen-Macaulay locus of $G(R)$ is Zariski-open. Suppose that the non-Cohen-Macaulay locus is not empty and let I be its defining ideal. It suffices to show that I is homogeneous.

If a is a unit of R, then there exists an R-automorphism of $G(R)$ which takes each form f of degree i to $a^i f$. Now let $f_0 + f_1 + \cdots + f_t$ be in I, where f_i is homogeneous of degree i. Choose units $a_0,\ldots,a_t$ in R with distinct residue classes modulo $\underline{m}$. Since I is invariant under every automorphism on $G(R)$, for $0 \leq j \leq t$,

$$\sum_{i=0}^{t} a_j^i f_i \in I$$

But $\det[a_j^i] = \pm \prod_{i<j} (a_i - a_j)$ is a unit in R. Therefore, each f_i is in I as desired.

3.10. <u>Theorem</u>. Let $(R,\underline{m})$ be a d dimensional local Cohen-Macaulay ring of embedding dimension $e(R) + d - 1$. Then $G(R)$ is Cohen-Macaulay.

<u>Proof</u>. We may assume that $R/\underline{m}$ is infinite. By (3.8), there exist elements $x_1,\ldots,x_d$ in $\underline{m}$ such that $\underline{m}^2 = \underline{x}\underline{m}$. We prove, by induction on d, that $\overline{x}_1,\ldots,\overline{x}_d$, the images of $x_1,\ldots,x_d$ in $G(R)$ form a regular sequence. By (3.9) this is sufficient to prove that $G(R)$ is Cohen-Macaulay. If $d = 1$, we have $\underline{m}^2 = x_1\underline{m}$. x_1 is not a zero divisor in $G(R)$, for $x_1 y \in \underline{m}^{t+1} = x_1\underline{m}^t$ implies that $y \in \underline{m}^t$ because x_1 is not a zero divisor in R. Assume that $d > 1$. We first check that x_1 is

not a zero divisor in $G(R)$. If $x_1 y \in \underline{m}^{t+1} = (x_1,\ldots,x_d)^t \underline{m}$, where $t > 1$, we must show that $y \in \underline{m}^t = (x_1,\ldots,x_d)^{t-1}\underline{m}$. $x_1 y = g(x_1,\ldots,x_d)x_1 + f(x_2,\ldots,x_d)$, where $g(x_1,\ldots,x_d)$ is a homogeneous polynomial of degree $t - 1$ in $x_1,\ldots,x_d$ with coefficients in $\underline{m}$ and $f(x_2,\ldots,x_d)$ is a homogeneous polynomial of degree t in $x_2,\ldots,x_d$ with coefficients in $\underline{m}$. Hence $hx_1 = f(x_2,\ldots,x_d)$ with $h = y - g(x_1,\ldots,x_d)$. Since $x_1,\ldots,x_d$ is a regular sequence in R, the associated graded ring of R with respect to the ideal $\underline{x} = (x_1,\ldots,x_d)$ is a polynomial ring in d variables over $R/\underline{x}$. It follows that $h \in (x_1,\ldots,x_d)^t$, (actually it follows, not quite so easily, that $h \in (x_2,\ldots,x_d)^t$), and therefore, that $y \in \underline{m}^t$. Pass to the Cohen-Macaulay ring R/x_1R. $e(R/x_1R) = e(R)$, $\dim R/x_1R = d - 1$ and $v(\underline{m}/x_1R) = v(\underline{m}) - 1$. The induction hypothesis applies to R/x_1R so that $x_2,\ldots,x_d$ form a regular sequence in $G(R/x_1R) = G(R)/x_1G(R)$. Hence $x_1, x_2,\ldots,x_d$ is a regular sequence in $G(R)$.

Examples. (1) If $t, X_1,\ldots,X_n$ are indeterminates, i is a positive integer and k is a field, then the local rings $k[[t^i, t^{i+1},\ldots,t^{2i-1}, X_1,\ldots,X_n]]$ satisfy the hypotheses of (3.7).

(2). It is also known that the local ring at a rational surface singularity satisfies the hypotheses of (3.7).

CHAPTER 3
LOCAL RINGS OF SMALL DIMENSION

The first attack on the problem of counting generators of ideals in a local ring $(R,\underline{m})$ is to look for properties of the ring R itself which will give information about numbers of generators of the ideals in R. From this point of view, the examples of Macaulay, mentioned in the introduction, are rather devastating because polynomial rings, being regular domains, are the best behaved rings of all. It turns out that the dimension of the ring is the crucial factor here.

In this chapter we look at local rings of small dimension. Some of the results described here are special cases of results for certain types of ideals in local rings of arbitrary dimension, cf. Chapter 5. We prefer, however, to treat the rings of small dimension first because the picture is so complete.

1. Local rings of dimension at most one.

The main results in this section consist of the addition of zero divisors to the result of Akizuki and Cohen mentioned in the introduction, and the identification of the bound in the case that the local ring is Cohen-Macaulay.

1.1. Theorem. Let $(R,\underline{m})$ be a local Cohen-Macaulay ring of dimension at most one. Let I be an ideal of R. Then

$$v(I) \leq e(R) - e(R/I), \text{ if } I \text{ has height } 0$$

and

$$v(I) \leq e(R), \text{ if } I \text{ has height } 1.$$

Proof. If $\dim R = 0$, then $e(R) = \lambda(R)$ so it is immediate that $v(I) \leq e(R) - e(R/I)$. Assume that $\dim R = 1$. By the standard trick of passing to $R(u) = R[u]_{\underline{m}R[u]}$, where u is an indeterminate, we may assume that $R/\underline{m}$ is an infinite field. Then, by (Chapter 1; 3.2), there is a nonzero divisor x in $\underline{m}$ such that x is a superficial element for R. The fact that x is a nonzero divisor means that xR is $\underline{m}$-primary, and then the condition for being superficial means that $x\underline{m}^t = \underline{m}^{t+1}$ for some $t > 0$. Suppose that I has height 1. Then,

$$(*) \qquad \lambda(I/xI) = \lambda(R/xR) = e(R).$$

The first equality follows from

$$\lambda(R/xR) = \lambda(R/xR) + \lambda(xR/xI) - \lambda(R/I) = \lambda(R/xI) - \lambda(R/I)$$

$$= \lambda(I/xI).$$

The second equality in (*) follows from the first by taking $I = \underline{m}^j$, for any integer $j \geq t$. The exact sequence

$$0 \to \underline{m}I/xI \to I/xI \to I/\underline{m}I \to 0,$$

gives $\lambda(I/\underline{m}I) = \lambda(I/xI) - \lambda(\underline{m}I/xI) = e(R) - \lambda(\underline{m}I/xI)$, so $v(I) \leq e(R)$.

Assume that I has height 0. By the Artin-Rees theorem, there is a positive integer s_0 such that $I \cap \underline{m}^s \subseteq I\underline{m}$, for $s \geq s_0$. $I/I\underline{m} = I/I\underline{m} + I \cap \underline{m}^s = I/I \cap I\underline{m} + \underline{m}^s \cong I + \underline{m}^s/I\underline{m} + \underline{m}^s$, for $s \geq s_0$. Thus,

$$e(R) \geq v(I + \underline{m}^s) = \lambda(I + \underline{m}^s/\underline{m}(I + \underline{m}^s)$$

$$= \lambda(I + \underline{m}^s/I + \underline{m}^{s+1}) + \lambda(I + \underline{m}^{s+1}/\underline{m}I + \underline{m}^{s+1}).$$

Then, by taking s large enough, it follows that

$$e(R) \geq e(R/I) + v(I).$$

Remarks. The result in (1.1) for ideals of height 1 can be found in [45], [47], [60] and is a special case of a theorem of Rees [56], cf. (Chapter 5; 4.4). The argument for ideals of height 0 is due to Kirby [40].

Note that it follows from the Hilbert-Samuel Theorem (Chapter 1; 2.1) that for some n, $v(\underline{m}^n) = e(R)$. By (Chapter 2; 2.3), if R is CM, then $v(\underline{m}^{e(R)-1}) = e(R)$.

1.2. Theorem. A local ring $(R,\underline{m})$ is of dimension at most one if and only if there is a non-negative integer n such that $v(I) \leq n$ for all ideals I in R.

Proof. It follows from the Hilbert-Samuel Theorem (Chapter 1; 2.1) that a bound on just the number of generators of powers of the maximal ideal $\underline{m}$ forces the dimension of the ring to be at most one.

Suppose, conversely, that R has dimension at most one. As the other cases follow from (1.1), we will assume that R is one dimensional and not CM. Just for the purpose of this proof we write $v(R) = n$ if R is a local ring with the property that every ideal of R can be generated by n elements and some ideal cannot be generated by n-1 elements. Let N be the nilradical of R and t the least integer such that $N^t = 0$. It is a straightforward induction on t to prove that

$$v(R) \leq e(R/N) \prod_{j=1}^{t-1} (1 + v(N^j)),$$

for the exact sequence

$$0 \to N^{t-1} \to R \to R/N^{t-1} \to 0$$

gives $v(R) \leq v(R/N^{t-1}) + v(N^{t-1})v(R/N^{t-1})$.

Example. The following example shows that the Cohen-Macaulay hypothesis is necessary in (1.1). Let k be a field and X and Y indeterminates. Let $R = k[[X,Y]]/(X^2,XY)$. Then $v(\underline{m}) = 2$ and $e(R) = 1$.

2. Local rings of dimension at most two.

We will see that two dimensional local rings retain as much of the boundedness properties of local rings of lower dimensions as can be expected in the light of the Hilbert-Samuel theorem. Clearly, there can be no bound on the number of generators of powers of the maximal ideal $\underline{m}$ of a two-dimensional local ring R. More generally, by forming quotients, we see that there can be no bound on the number of generators of ideals which have $\underline{m}$ as an associated prime. However, there is a bound on the number of generators of all other ideals.

2.1. Theorem. A local ring $(R,\underline{m})$ has dimension at most two if and only if there is a bound on the number of generators of all ideals I such that $\underline{m}$ is not an associated prime of I.

Proof. Assume that $(R,\underline{m})$ is a local ring with the property that there is a bound on the number of generators of all ideals I such that $\underline{m}$ is not an associated prime of I. If R has dimension > 2, let P be a height 2 prime, $P \neq \underline{m}$. Consider the symbolic powers $P^{(n)}$ of P, where $P^{(n)} = P^nR_P \cap R$. By hypothesis, these ideals admit a bounded number of generators. However, this implies that R_P has dimension at most one, a contradiction.

For the converse, by (1,2), we may assume that $(R,\underline{m})$ is two dimensional. Let x,y be a system of parameters for R. Suppose that I is an ideal having no $\underline{m}$-primary component. Then $(x,y)R$ is not in any prime belonging to I. Let u be an indeterminate and pass to the ring $R(u) = R[u]_{\underline{m}R[u]}$. Let $f = x + yu \in R[u]$. Then $fR(u) \cap IR(u) = fIR(u)$, because the primes belonging to $IR(u)$ are the extensions to $R(u)$ of the primes belonging to I in R. Now pass to the 1-dimensional local ring $\overline{R(u)} = R(u)/fR(u)$ and to the ideal $\overline{IR(u)} =$

$(I,f)R(u)/fR(u) \cong IR(u)/fR(u) \cap IR(u) = IR(u)/fIR(u)$. By (1,2), there is a positive integer n such that $v(IR(u)) \leq n$. But

$$v(I) = v(IR(u)) = v(\overline{IR(u)}),$$

the last equality by Nakayama's lemma.

Remark. It is an open question whether a bound on the number of generators of prime ideals implies that the ring has dimension at most two.

Next we consider families of mixed ideals having an $\underline{m}$-primary component generated by a fixed number j of elements. We show that any member of such a family admits a bound on its number of generators where the bound depends on j. This will follow from the fact, which we prove now, that any two-dimensional local ring is uniformly coherent (cf. Chapter 1; 8).

2.2. Theorem. Any two-dimensional local ring is uniformly coherent.

Proof. The proof involves reduction to the case of a two-dimensional complete local domain R. For such a domain is by [51; (31.6)] a finitely generated module over a complete regular local ring S. But S is uniformly coherent, for the kernel of any nonzero homomorphism $f: S^n \to S$ is a free module of rank $n - 1$ since S has global homological dimension 2. Thus an application of (Chapter 1; 8.5) gives that R is uniformly coherent.

We make the reduction as follows. It is clear that we may assume that R is complete. By (Chapter 1; 8.6) we may pass from R to R/N, where N is the nilradical of R, and then from R/N to $R/P_1 \oplus \cdots \oplus R/P_t$, where $P_1,\ldots,P_t$ are the minimal primes of R. It is clear that any finite direct sum of uniformly coherent rings is uniformly coherent and this finishes the reduction.

Remark. In [54], Quentel gives a neat proof that there

is a bound on the number of generators of prime ideals in any uniformly coherent Noetherian ring.

The next lemma clarifies the connection between the notion of uniform coherence and bounds for numbers of generators of intersections of ideals.

2.3. Lemma. Let $(R,\underline{m})$ be a uniformly coherent local rings with ideals J and L. Then, $v(J \cap L) \leq \varphi(v(J) + v(L))$.

Proof. If we apply the snake lemma to the commutative diagram:

$$\begin{array}{ccccccccc}
0 \to \ker f_1 \oplus \ker f_2 & \to & R^{v(J) + v(L)} & \xrightarrow{f_1 \oplus f_2} & J \oplus L & \to 0 \\
\downarrow & & \downarrow & & \downarrow & \\
0 \longrightarrow \ker f & \to & R^{v(J) + v(L)} & \xrightarrow{f} & J + L & \to 0 \\
& & & & \downarrow & \\
& & & & 0 &
\end{array}$$

where $f_1\colon R^{v(J)} \to J$, $f_2\colon R^{v(L)} \to L$ and $f\colon R^{v(J) + v(L)} \to J + L$ are the obvious maps, we obtain the isomorphism
$J \cap L \cong \ker f/\ker f_1 \oplus \ker f_2$.

2.4. Theorem. Let $(R,\underline{m})$ be a two dimensional local ring. Let I be an ideal of height at most one and write $I = J \cap L$, where J is $\underline{m}$-primary and L has no $\underline{m}$-primary component. Then there is a positive integer $N(j)$, depending only on $j = v(J)$, such that $v(I) \leq N(j)$.

Proof. By (2.2), R is uniformly coherent. By (2.3), $v(I) = v(J \cap L) \leq \varphi(v(J) + v(L))$. By (2.1), $v(L) \leq n$, where n is a positive integer depending only on R. Thus $v(I) \leq \varphi(v(J) + n) = N(j)$, where $j = v(J)$.

Examples. (1). If $(R,\underline{m})$ is a two-dimensional regular local ring, then $N(j) = j$. For let J be an $\underline{m}$-primary ideal with $v(J) = j$. Let z be any nonzero element of R. If

$f: R^{j+1} \to (J,z) \to 0$ is a nonzero homomorphism, then ker f is a free module of rank j. Hence $N(j) \leq j$. Let $\underline{m} = (x,y)$ and let $J = (x^j + y^{j+1}, x^{j-1}y^2, \ldots, xy^j)$. Then $J \cap (x) = (x^{j-1}y^2, \ldots, xy^j) + (x^j + y^{j+1}) \cap (x) = (x^{j+1} + y^{j+1}x, x^{j-1}y^2, \ldots, xy^j)$.

(2). Let $(R,\underline{m})$ be a two-dimensional CM local ring containing a field. Let J and L be ideals of R. Then $v(J \cap L) \leq e(R)(j + \ell - 1)$, where $j = v(J)$, $\ell = v(L)$. For we may assume that R is complete with infinite residue field. If $k \cong R/\underline{m}$ and $k \subset R$, then there is a s.o.p. x, y in R such that R is a free module of rank e(R) over k[[x,y]]. From the exact sequence

$$0 \to \ker f \to R^{j+\ell} \xrightarrow{f} (J,L) \to 0$$

we see that ker f is a free k[[x,y]]-module of rank $e(R)(j + \ell - 1)$. Hence, $v(J \cap L) \leq e(R)(j + \ell - 1)$. For example, $R = k[[x^2,xy,y^2]]$ is a CM local ring of multiplicity 2. Let $J = (x^4,y^4)$ and $L = (xy)$. Then $J \cap L = (x^5y, xy^5, x^4y^2, x^2y^4)$ and so $v(J \cap L) = e(R)(j + \ell - 1)$.

If we assume that the two-dimensional local ring is Cohen-Macaulay then much more is known. Boratynski and Eisenbud [9] have given very precise bounds in this case. The proof we give below of one of the results of Boratynski and Eisenbud was inspired by Rees' paper [56]. First, we need a lemma.

We will say that an ideal I in a local ring $(R,\underline{m})$ of arbitrary dimension d has n irreducible $\underline{m}$-primary components if the $\underline{m}$-primary component J of I can be written as an irredundant intersection $J = J_1 \cap \ldots \cap J_n$, where the J_i are irreducible ideals.

2.5. Lemma. With the notation as above,

$$n = \dim_{R/\underline{m}} \mathrm{Hom}(R/\underline{m}, R/I).$$

Proof. We may assume that I has an $\underline{m}$-primary component J; otherwise n = 0. $\mathrm{Hom}_R(R/\underline{m}, R/I) \cong \mathrm{Hom}_R(R/\underline{m}, R/J) \cong (J : \underline{m})/J$.

Passing to the ring R/J, we may assume that R has dimension 0 and that $0 = J_1 \cap \ldots \cap J_n$ is an irredundant intersection of irreducible ideals. Let E(A) denote the injective envelope of the R-module A. If $t = \dim_{R/\underline{m}}((0 : \underline{m}))$, then $E((0 : \underline{m})) = E(R/\underline{m})^t$. But $E((0 : \underline{m})) = E(R) = E(R/J_1) \oplus \ldots \oplus E(R/J_n) = E(R/\underline{m})^n$. Thus n = t.

2.6. Theorem. Let $(R,\underline{m})$ be a two dimensional local Cohen-Macaulay ring. Let I be an ideal having n irreducible $\underline{m}$-primary components. Then

$$v(I) \leq (n + 1)e(R),$$

with strict inequality if height I = 0.

Proof. We may assume that $R/\underline{m}$ is infinite. Let $\underline{x} = x_1,x_2$ be elements of R such that $e(R) = e(\underline{x}R) = \lambda(R/\underline{x}R)$, cf. (Chapter 1; 3.3) and (Chapter 1; 3.4). The exact sequence $0 \to I \to R \to R/I \to 0$ gives, upon tensoring with $R/\underline{x}R$, the exact sequence

$$0 \to \mathrm{Tor}_1^R(R/\underline{x}R), R/I) \to I/\underline{x}I \to R/\underline{x}R \to R/(\underline{x},I) \to 0.$$

So we have,

$$\lambda(I/\underline{x}I) = \lambda(R/\underline{x}R) + \lambda(\mathrm{Tor}_1^R(R/\underline{x}R,R/I)) - \lambda(R/(\underline{x},I)).$$

Since $\underline{x}$ is an R-sequence, $\mathrm{Tor}_i^R(R/\underline{x}R,R/I) = H_i(K(\underline{x}; R/I))$. By (Chapter 1; 4.3)

$$\sum_{i=0}^{2} (-1)^i \mathrm{Tor}_i^R(R/\underline{x}R,R/I) = e(\underline{x}R; R/I).$$

By (Chapter 1; 2),

$$e(\underline{x}R;R/I) = \begin{cases} 0, & \text{if height } I > 0 \\ \lambda(R/(\underline{x},I), & \text{if height } I = 0. \end{cases}$$

Thus, if height I = 0, $\lambda(I/\underline{x}I) = e(R) + \lambda(\mathrm{Tor}_2^R(R/\underline{x}R,R/I))$ -

$\lambda(R/(\underline{x},I))$. If height $I \geq 1$, $\lambda(I/\underline{x}I) = e(R) + \lambda(\mathrm{Tor}_2^R(R/\underline{x}R,R/I))$. But $\mathrm{Tor}_2^R(R/\underline{x}R,R/I) \cong (I; \underline{x}R)/I$, so it is sufficient to prove that $\lambda((I: \underline{x}R)/I) \leq ne(R)$.

Let $I = L \cap J_1 \cap \ldots \cap J_n$ be an irredundant intersection of ideals where L has no $\underline{m}$-primary component and $J_1,\ldots,J_n$ are $\underline{m}$-primary irreducible ideals. Then $(I: \underline{x}R) = (L : \underline{x}R) \cap (J_1 : \underline{x}R) \cap \ldots \cap (J_n : \underline{x}R) = L \cap (J_1 : \underline{x}R) \cap \ldots \cap (J_n : \underline{x}R)$. Now $(I : \underline{x}R)/I$ maps isomorphically into $(J_1 : \underline{x}R)/J_1 \oplus \ldots \oplus (J_n : \underline{x}R)/J_n$ by the map

$$r \to (r + J_1/J_1,\ldots,r + J_n/J_n).$$

Hence

$$\lambda((I : \underline{x}R)/I) \leq \sum_{i=1}^{n} \lambda((J_i : \underline{x}R)/J_i) = \sum_{i=1}^{n} \lambda(R/(J_i,\underline{x}) \leq ne(R).$$

The middle equality follows from the fact that R/J_i is zero dimensional Gorenstein, and the last inequality from the fact that $\lambda(R/\underline{x}R) = e(R)$.

Examples. (1). Let $R = k[[X,Y,Z]]/(X^2 - Y^2)$. Let x,y,z denote the images of X, Y, Z in R. Let $I = (y^2-z^2,xy,yz,zx)R$. I is an irreducible ideal of R. (R/I is a zero dimensional Gorenstein ring for $(0 : \underline{m}/I) = (x^2 + I/I)R/I \cong R/\underline{m}$.) $e(R) = 2$ and $v(I) = (1 + 1)2$, so that the bound given in (2.6) is attained in this case.

(2). In [9] examples are given of 2 dimensional CM rings $(R_e,\underline{m}_e)$ and irreducible $\underline{m}_e$-primary ideals I_e such that $v(I_e) = 2e$, where e is the multiplicity of R_e. The examples are constructed as follows. Take the ideals H_n, constructed by Buchsbaum and Eisenbud, which are defined following (Chapter 5; 1.7). Let $n = 2e + 1$. The $2e + 1$ generators of H_n are homogeneous forms of $k[[X,Y,Z]]$ of degree e. Let $R_e = k[[X,Y,Z]]/hk[[X,Y,Z]]$, where h is one of the generators of H_n and define $I_e = H_n/hk[[X,Y,Z]]$.

(3). In (Chapter 5; 2.2) we will see that a different bound can be given for $\underline{m}$-primary ideals in a two dimensional local CM ring $(R,\underline{m})$. If I is such an $\underline{m}$-primary ideal, then

$$v(I) \leq te(R) + 1,$$

where t is the least integer such that $\underline{m}^t \subseteq I$. For the example given above this bound is not as good as the one in (2.6), but sometimes, as in the example to follow, this bound is better. Let $R = k[[x^2,xy,y^2]]$, where k is a field and x and y are indeterminates. $e(R) = 2$. If $I = \underline{m}^t = (x^2,xy,y^2)^t$, then $v(I) = 2t + 1 < 2(t + 1)$.

3. Unboundedness in dimension greater than two.

The famous examples of Macaulay show that the boundedness which is so apparent in local rings of dimension at most two all but disappears in dimension greater than two. The reader is referred to [2] and [29] for expositions of Macaulay's primes. Here we will give very brief descriptions of some other examples of unboundedness.

Macaulay's primes are not analytically irreducible: when extended to the power series ring in three variables they are no longer prime. Moh [50] has constructed examples of analytically irreducible prime ideals P_n in $K[X,Y,Z]$, K a field of characteristic 0, such that P_n needs at least n generators in $K[[X,Y,Z]]$. The primes P_n are easy to describe but the proof that P_n needs at least n generators requires quite a bit of computation. Let n be an odd positive integer and $m = (n + 1)/2$. Let ℓ be an integer $> n(n + 1)m$ with $(\ell,m) = 1$. Let X, Y, Z and t be indeterminates and let ρ be the homomorphism $K[[X,Y,Z]] \to K[[t]]$ defined by:

$$\rho(X) = t^{nm} + t^{nm+\ell}$$
$$\rho(Y) = t^{(n+1)m}$$
$$\rho(Z) = t^{(n+2)m}.$$

If $P_n = \ker \rho$, then $v(P_n) \geq n$.

To fully appreciate the definition of the P_n a little background information may be helpful. Let S be a numerical semigroup, i.e., a subsemigroup of the nonnegative integers $\mathbb{N}$ such that the generators of S have greatest common divisor 1. Numerical semigroups arise, for example, as value semigroups of one dimensional analytically irreducible local rings. In particular, the rings $k[[t^{a_1},t^{a_2},\ldots,t^{a_n}]]$, where k is any field and $a_1,\ldots,a_n$ form a minimal set of generators of a numerical semigroup S, provide a wealth of examples in local ring theory.

S is a numerical semigroup if and only if there is a nonnegative integer s such that $s + \mathbb{N} \subset S$. The smallest such integer c is called the conductor of S. S is said to be symmetric if the number of elements in S that are less than c is $c/2$; we write

$$\operatorname{card}(\{0,1,\ldots,c-1\} \cap S) = c/2.$$

Kunz [42] proved that a one dimensional analytically irreducible local ring $(R,\underline{m})$ with residue field equal to that of its integral closure is Gorenstein if and only if the value semigroup of R is symmetric. So, for example, the rings $k[[t^a,t^b]]$ with $(a,b) = 1$ are Gorenstein, whereas the rings $k[[t^a,t^{a+1},t^{a+2},\ldots,t^{2a-1}]]$ are Gorenstein if and only if $a = 2$.

Of particular interest in relation to Moh's examples is the rèsult of Herzog [32] that if S is a numerical semigroup generated say by a_1,a_2,a_3, then the kernel of the homomorphism

$$\mu\colon\ k[X,Y,Z] \to k[t^{a_1},t^{a_2},t^{a_3}],$$

defined by $\mu(X) = t^{a_1}$, $\mu(Y) = t^{a_2}$, $\mu(Z) = t^{a_3}$ can be generated by three elements, and that two elements suffice if S is symmetric. (The last part of the statement follows from a more general result due to Serre, cf (Chapter 5; 1.2), because in this case the kernel is a Gorenstein ideal.) On the other hand, Brezinsky [12] has shown that if S is minimally generated by $a_1, a_2, \ldots, a_r$ with $r > 3$ then there is no bound for the number of generators of primes arising as kernels of homomorphisms

$$k[X_1, \ldots, X_r] \to k[t^{a_1}, \ldots, t^{a_r}],$$

defined by $X_i \to t^{a_i}$.

We briefly indicate the steps involved in Moh's construction. We assume that K is a field of characteristic 0. First, let α be the homomorphism $K[[X,Y,Z]] \to K[[X,Y,Z]]$ defined by $\alpha(X) = X^{a_1}$, $\alpha(Y) = Y^{a_2}$, $\alpha(Z) = Z^{a_3}$, where $a_1 \leq a_2 \leq a_3$ are positive integers. Let f be an element of $K[[X,Y,Z]]$. Define the α-order of f to be the order of $\alpha(f)$, and the α-leading form of f to be α^{-1} (leading form of $\alpha(f)$). f is α-homogeneous if f = α-leading form of f. Let W_r be the vector space of all α-homogeneous power series of α-order r. Let I be any ideal of $K[[X,Y,Z]]$ and

$$V_r(I) = W_r \cap \{\alpha\text{-leading forms of elements of } I\} \cup \{0\}.$$

Set $s(I) = \min\{\alpha\text{-order } f \mid f \in I\}$.

Step 1. $v(I) \geq \sum_{r=s(I)}^{s(I)+a_1-1} \dim V_r(I)$.

Now we return to the actual definition of the primes P_n. Recall that we set n to be an odd positive integer, $m = (n + 1)/2$ and ℓ an integer $> n(n + 1)m$ with $(\ell, m) = 1$. Then $P_n = P$ was defined to be the kernel of the homomorphism $K[[X,Y,Z]] \xrightarrow{\rho} K[[t]]$ defined by $\rho(X) = t^{nm} + t^{nm+\ell}$, $\rho(Y) = t^{(n+1)m}$ and $\rho(Z) = t^{(n+2)m}$. In the definition of α above, set $a_1 = n$, $a_2 = n + 1$ and $a_3 = n + 2$.

Step 2. Computation of $\dim W_r$.

(a) If $r < n(n + 1)$, then $\dim W_r \leq m$.

(b) If $r < (n + 1)(n + 2)$ and $n(n + j) \leq r < n(n+j+1)$, then $\dim W_r = m + j$. Moh proves (a) and (b) by showing that if $r < (n+1)(n+2)$, then $\dim W_r = \mathrm{Card}(S \cap \{r + nZ\} \cap \{0,1,\ldots,r\})$, where S is the semigroup generated by $n+1$ and $n+2$.

Step 3. Computation of $\dim V_r(P)$.

(c) If $r < n(n+1)$, then $\dim V_r(P) = 0$.
(d) If $n(n+1) \leq r \leq n(n+2) - 1$, then $\dim V_r(P) = 1$.

Moh proves (c) and (d) using the linear independence of certain binomial vectors. Let p be a nonnegative integer and q a positive integer. Define

$$b_{p,q} = (\tbinom{p}{0}, \tbinom{p}{1}, \ldots, \tbinom{p}{q-1}).$$

$b_{p,q} \in Z^q \subseteq K^q$. If $0 \leq t_1 < t_2 < \cdots < t_q$ are integers, the vectors $b_{t_i,q}$ are linearly independent over K. Define a map $\varphi\colon W_r \to K^m$ by $\varphi(X^\alpha Y^\beta Z^\gamma) = b_{\alpha,m}$. Then φ is injective if $r < n(n+1)$; φ is surjective if $r \geq n(n+1)$ and $\ker \varphi = V_r(P)$.

Remark. Once Moh's primes are constructed in $K[[X,Y,Z]]$ with K of characteristic 0, then one can see that they also exist in $k[[X,Y,Z]]$ for k a field of any characteristic. Assume that k has characteristic $p > 0$. Fix n as above and consider the mapping ρ where we take $K = \mathbb{Q}$, rationals. Analogous maps $\tilde{\rho}$ and $\bar{\rho}$ are defined to yield the following commutative diagram:

$$\begin{array}{ccc}
\mathbb{Z}/p\mathbb{Z}[[X,Y,Z]] & \xrightarrow{\bar{\rho}} & \mathbb{Z}/p\mathbb{Z}[[t]] \\
\uparrow & & \uparrow \\
\mathbb{Z}_p[[X,Y,Z]] & \xrightarrow{\tilde{\rho}} & \mathbb{Z}_p[[t]] \\
\downarrow & & \downarrow \\
\mathbb{Q}[[X,Y,Z]] & \xrightarrow{\rho} & \mathbb{Q}[[t]]
\end{array}$$

Let $\tilde{P}_n = \ker \tilde{\rho}$ and $\bar{P}_n = \ker \bar{\rho}$. Clearly $\tilde{P}_n$ needs at least as many generators as P_n. $\overline{P}_n = (\tilde{P}_n, p)\mathbb{Z}_p[[X,Y,Z]]/p\mathbb{Z}_p[[X,Y,Z]] = \tilde{P}_n\mathbb{Z}_p[[X,Y,Z]]/p\tilde{P}_n\mathbb{Z}_p[[X,Y,Z]]$, since $p \nmid \tilde{P}_n$. Thus it follows by Nakayama's lemma that $\overline{P}_n$ needs at many generators as P_n. Now extend $\overline{\rho}$ to ρ_k: $k[[X,Y,Z]] \to k[[t]]$. If $Q_n = \ker \rho_k$, Q_n needs at least n generators. Thus it immediately follows that "Moh's primes" exist in any 3 dimensional regular local ring which contains a coefficient field.

CHAPTER 4
IDEALS GENERATED BY R-SEQUENCES

The Krull principal ideal theorem gives a lower bound for the number of generators of an ideal. One of the first - and perhaps the hardest - questions to ask is: "When can an ideal I in a local ring $(R,\underline{m})$ be generated exactly by height I generators?" If we make the assumption that R is Cohen-Macaulay, we are asking: "When can I be generated by an R-sequence?" From the geometric standpoint, this question is the most interesting because it is the question of when a variety is a complete intersection at a given point.

Most of the "older" (say, up through 1965) characterizations of ideals generated by R-sequences are well known, and can be found, for example in [61]. Some of the techniques used more recently are quite geometric and are beyond the scope of these notes. We note, in particular, the work of Hartshorne and Ogus [31] on local complete intersections in which duality theory plays a big role. Here we will give two results which have some of the flavor of these geometric techniques.

Note. We say that the local ring $(S,\underline{m})$ is a (local) complete intersection if $S = R/I$, where R is a regular local ring and I is generated by a R-sequence.

1. Use of the module of differentials.

For the following result, (1.1), due to Ferrand [28] and Vasconcelos [68], the module of Kahler differentials is brought into play. The setting is geometric. Given the local ring $(R,\underline{m})$ of an affine variety at a point, then (1.1) characterizes $(R,\underline{m})$ as a complete intersection in terms of the projective dimension of the module of differentials.

1.1. Theorem. Let k be a field. Let $P' \subset P$ be two prime ideals in the polynomial ring $k[X_1,\ldots,X_n]$. Let $R = k[X_1,\ldots,X_n]_P$ and $S = R/Q$, where $Q = (P')_P$. Assume that K, the quotient field of S, is separable over k. Then Q is generated by an R-sequence if and only if $\Omega_{S/k}$, the module of differentials of S over k, has projective dimension at most one.

Remark. It is well known, cf. [48], that - in the situation described in (1.1) - S is regular if and only if $\Omega_{S/k}$ is S-free.

For the proof of (1.1) we will use a result of Vasconcelos which is a variation of the well known characterization of R-sequences due to Ferrand [28] and Vasconcelos [68], namely that if I is an ideal of finite projective dimension in a local ring $(R,\underline{m})$, then I can be generated by an R-sequence if and only if I/I^2 is a free R/I-module.

1.2. Proposition. Let I be an ideal of finite projective dimension in the local ring $(R,\underline{m})$. Assume that $I/I^2 \cong (R/I)^r \oplus T$, where T is an R/I-module. If $T = 0$, I is generated by an R-sequence. If $T \neq 0$, I contains an R-sequence of length $r + 1$.

Proof. The proof is by induction on r. By (Chapter 1; 4.7), I contains a nonzero divisor, so we may assume that $r > 0$. $I/I^2 \cong (R/I)^r \oplus T$ implies that there are ideals J and L in R such that $J + L = I$, $J \cap L = I^2$ and $J/I^2 \cong (R/I)^r$. Since I contains a nonzero divisor, we may, by the standard prime avoidance argument, choose a set $x_1,x_2,\ldots,x_r$ of elements of I which map onto a free basis of the R/I-module J/I^2, where x_1 is a nonzero divisor in $I \setminus \underline{m}I$. This means that any relation $\Sigma a_i x_i \in I^2$ implies all $a_i \in I$. Pass to the ring $R^* = R/x_1R$ and the ideal $I^* = I/x_1R$. We have $I^*/(I^*)^2 \cong (R^*/I^*)^{r-1} \oplus T^*$. In order to apply the induction hypothesis we need that I^* has finite projective dimension in R^*. But I/x_1I does have finite projective dimension over R^*

so it suffices to show that I^* is a direct summand of I/x_1I. We show that $I/x_1I \cong x_1R/x_1I \oplus I^*$. It is clear that $I/x_1I = x_1R/x_1I + I^*$. Suppose that $z \in x_1R \cap (x_2,\ldots,x_r,L)$. Then $z = x_1a = a_2x_2 + \cdots + a_rx_r + \ell$. Now $\ell \in J \cap L = I^2$, so $a \in I$, as required. Thus we may use induction to complete the proof.

Proof of (1.1). We have the exact sequence of S-modules

$$Q/Q^2 \to \Omega_{R/k} \otimes_R S \to \Omega_{S/k} \to 0.$$

If $\Omega_{S/k}$ has projective dimension at most one, then Q/Q^2 maps onto F, a free S-submodule of $\Omega_{R/k} \otimes_R S$. We have the exact sequence

$$0 \to F \otimes_S K \to \Omega_{R/k} \otimes_R S \otimes_S K \to \Omega_{S/k} \otimes_S K \to 0.$$

Using the hypothesis of separability, we can count ranks as follows:

$$\Omega_{R/k} \otimes_R S \otimes_S K \cong R^n \otimes_R S \otimes_S K \cong K^n \text{ and } \Omega_{S/k} \otimes_S K \cong \Omega_{K/k} \cong K^t,$$

where t is the transcendence degree of K over k. Thus F is a free S-module of rank $n - t$. But $n - t$ is the height of Q, so it follows from (1.2) that Q is generated by an R-sequence.

Conversely, assume that Q is generated by an R-sequence. Then Q/Q^2 is free of rank $n - t$ over S. Thus, if C is the kernel of the map $Q/Q^2 \to \Omega_{R/k} \otimes_R S$, then, counting ranks as above, we have $C \otimes_S K = 0$. But C, being a submodule of a free module, is torsion free; hence $C = 0$.

2. Use of the punctured spectrum of a local ring.

In this section we will look at the local version of the theorem of Szpiro [66] on complete intersections of codimension two in P^n. The proof gives a slight indication of the use of the punctured spectrum of a local ring. (If $(R,\underline{m})$ is a local ring, the punctured spectrum of R is the space

Spec$R$ - $\{\underline{m}\}$. In characteristic zero, Hartshorne and Ogus [31] have used duality theory of the punctured spectrum in a very powerful way to obtain results on complete intersections.

2.1. Lemma. Let $(R,\underline{m})$ be a local ring and I an ideal of R. Suppose that the sequence

$$0 \to R^n \xrightarrow{\alpha} R^{n+1} \to I \to 0$$

is exact. Then I can be generated by $t < n + 1$ elements if and only if the minors of $[\alpha]$ of order $n + 1 - t$ generate R.

Proof. If I is minimally generated by $n + 1$ elements, then $\alpha(R^n) \subseteq \underline{m}R^{n+1}$ so that the minors of $[\alpha]$ of every order $\leq n$ are ideals properly contained in R.

Suppose that I can be generated by $t < n + 1$ elements. Then the entries of $[\alpha]$ do not all lie in $\underline{m}$. If $\ell = n + 1 - t$, the previous remark takes care of the case $\ell = 1$. Assume $\ell > 1$. Then the matrix $[\alpha]$ is equivalent to a matrix of the form

$$\begin{bmatrix} 1 & 0 \\ 0 & [\beta] \end{bmatrix},$$

where $[\beta]$ is an $n \times n - 1$ matrix corresponding to a map β: $R^{n-1} \to R^n$, with cokernel $\beta = I$, and the ideal generated by the minors of order j of $[\alpha]$ is the same as the ideal generated by the minors of order $j - 1$ of $[\beta]$. So the proof can be completed by induction.

Here is Szpiro's result.

2.2. Theorem. Let $(R,\underline{m})$ be an equicharacteristic regular local ring of dimension $d \geq 7$. Let I be an ideal of height 2 such that:

(i) R/I is Cohen-Macaulay

(ii) IR_P is generated by 2 elements for all

$P \in \mathrm{Spec}R - \{\underline{m}\}$. Then I is generated by 2 elements.

Since we are interested in estimates of numbers of generators of arbitrary ideals, we generalize (2.2) a little bit as follows.

2.3. Theorem. Let $(R,\underline{m})$ be an equicharacteristic regular local ring of dimension d. Let I be an ideal of height 2 such that:

(i) R/I is Cohen-Macaulay

(ii) IR_P is generated by t elements for all $P \in \mathrm{Spec}R - \{\underline{m}\}$. Then, if $d > t(t + 1)$, I is generated by t elements.

Proof. We may assume that $R = k[[X_1,\ldots,X_d]]$, k a field. By (Chapter 1; 4.6), the hypotheses on I imply that projective dimension $I = 1$. Suppose that

$$0 \to R^n \xrightarrow{\alpha} R^{n+1} \to I \to 0$$

is a minimal free resolution of I with $n \geq t$. Let J be the ideal of R generated by the minors of $[\alpha]$ of order $n - t + 1$. By (2.1), $JR_P = R_P$, for all $P \in \mathrm{Spec}R - \{\underline{m}\}$. Thus J is an $\underline{m}$-primary ideal. Let $[\alpha] = [\alpha_{ij}]$, $1 \leq i \leq n + 1$, $1 \leq j \leq n$, and let $S = k[[Y_{ij}]]$, $1 \leq i \leq n + 1$, $1 \leq j \leq n$, where the Y_{ij} are indeterminates. Let φ be the homomorphism of S into R defined by $Y_{ij} \to \alpha_{ij}$. φ takes the maximal ideal of S onto an $\underline{m}$-primary ideal of R. Thus R is integral over $T = k[[\alpha_{ij}]]$ and $\dim T = d$. Let L be the ideal generated in S by the minors of the matrix $[Y_{ij}]$ of order $n - t + 1$. We are going to compute height L in two different ways and arrive at a contradiction to the hypothesis that $d > t(t + 1)$. By (Chapter 1; 7.1), height $L \leq (n+1-(n-t+1)+1)(n-(n-t+1)+1) = (t+1)(t)$. Let $\mathscr{L} = \ker \varphi$. Then, height $\mathscr{L} = (n+1)n - d$. By (Chapter 1; 7.2), height $(\mathscr{L} + L) \leq$ height $\mathscr{L}$ + height L. But $\mathscr{L} + L$ is primary for the maximal ideal of S. For if we let J_0 be the ideal of T generated by the minors of order

$n + t - 1$, then $S/\mathscr{L} + L \cong T/J_0$, and J_0 is primary for the maximal ideal of T. Thus $(t + 1)t \geq$ height $L \geq (n + 1)n - (n + 1)n + d$, which is the desired contradiction.

In [66], Szpiro gives examples to show that (2.2) fails for $d < 7$. One of his examples is as follows. Let k be a field and let $(R,\underline{m}) = (k[[X_0,X_1,X_2,X_3]], (X_0,X_1,X_2,X_3))$. Let I be the ideal generated by $X_0X_2 - X_1^2$, $X_0X_3-X_1X_2$, $X_1X_3 - X_2^2$. Then a minimal resolution for I is

$$0 \to R^2 \xrightarrow{\begin{bmatrix} X_0 & X_1 \\ X_1 & X_2 \\ X_2 & X_3 \end{bmatrix}} R^3 \to I \to 0.$$

By (2.1), I_P is generated by two elements for all $P \in \operatorname{Spec} R - \{m\}$, but I is minimally generated by three elements.

CHAPTER 5
HOW R/I DETERMINES v(I)

Macaulay's and Moh's primes illustrate that it is not, in general, sufficient to look at properties of the ring R itself to get information about numbers of generators of ideals in R. Other "variables" will have to be introduced. For example, each prime of Macaulay or Moh represents an affine or algebroid curve, and we will see that the number of generators of the prime is bounded by the multiplicity of the curve. In this chapter we will study how properties of R/I determine $v(I)$.

Often, but not always, the properties which we assume about R/I to get information about $v(I)$ will be "at least as good" as the properties which we assume for R, where, beginning with "best", the hierarchy of "goodness" is: regular, complete intersection, Gorenstein and Cohen-Macaulay.

Regular quotients were discussed in Chapter 2 and complete intersections in Chapter 4. In this chapter we look at Gorenstein quotients and Cohen-Macaulay quotients.

1. Gorenstein ideals.

1.1. Definition. Let $(R,\underline{m})$ be a regular local ring. An ideal I of R is a Gorenstein ideal if R/I is a Gorenstein ring.

The "classical" result about numbers of generators of Gorenstein ideals is due to Serre.

1.2. Theorem. Let I be a Gorenstein ideal of height two in a regular local ring $(R,\underline{m})$. I can be generated by two elements.

We use the following lemma for the proof of (1.2).

1.3. Lemma. Let $(R,\underline{m})$ be a regular local ring and I an ideal of height two. If I has projective dimension one, then

$$v(I) = v(\mathrm{Ext}_R^2(R/I,R)) + 1.$$

Proof. Let $v(I) = n$ and let

$$(*) \qquad 0 \to R^{n-1} \to R^n \to R \to R/I \to 0$$

be a minimal resolution of R/I. By (Chapter 1; 4.6), R/I is CM. Thus $\mathrm{Ext}_R^i(R/I,R) = 0$, except for $i = 2$, cf. (Chapter 1; 5.1). Thus, if we apply $\mathrm{Hom}_R(-,R)$ to $(*)$, we get a minimal free resolution of $\mathrm{Ext}_R^2(R/I,R)$

$$0 \to R \to R^n \to R^{n-1} \to \mathrm{Ext}_R^2(R/I,R) \to 0.$$

Proof of (1.2). By Chapter 1 (5.2), $\mathrm{Ext}(R/I,R) \cong R/I$. Thus, by (1.3), $v(I) = 2$.

Example. The classical example showing that Serre's result does not generalize to height three Gorenstein ideals is as follows. Let k be a field. Let $(R,\underline{m}) = (k[[X,Y,Z]]/(X^2-Y^2,Y^2-Z^2,XY,YZ,ZX), (x,y,z))$. Then R is Gorenstein for the annihilator of $\underline{m}$ is x^2R which is one dimensional over $R/\underline{m}$, but R is not a complete intersection.

The structure of height three Gorenstein ideals has been discovered by Buchsbaum and Eisenbud [14]. They have characterized these ideals as the Pfaffians of certain alternating matrices. As a corollary, it follows that the minimal number of generators of such an ideal must be odd. This last result was also obtained by Watanabe [69].

The Buchsbaum-Eisenbud characterization requires a certain amount of preliminary material, beginning with the definition of Pfaffian.

1.4. Definition. Let R be any commutative ring and let F be a finitely generated free R-module. A homomorphism

$f\colon F \to F^* = \mathrm{Hom}_R(F,R)$ is alternating if, with respect to some basis of F and corresponding dual basis of F^*, the matrix of f is skew-symmetric and all diagonal entries are zero. Given such an alternating map f, fix a basis of F and a dual basis for F^*. If the rank of F is even, then there is an element $\mathrm{Pf}(f) \in R$ called the Pfaffian of f which is a polynomial function of the entries of $[f]$ such that $\det([f]) = (\mathrm{Pf}(f))^2$, cf. [3; p. 141].

In general, given a free summand G of F generated by an even number $2j$ of the basis elements of F, we say that the Pfaffian of the composite homomorphism:

$G \xrightarrow{i} F \xrightarrow{f} F^* \xrightarrow{i^*} G^*$, where i is the inclusion, is a Pfaffian of f of order $2j$. Let $\mathrm{Pf}_{2j}(f)$ be the ideal of R generated by the Pfaffians of f of order $2j$.

Buchsbaum and Eisenbud prove that, for any odd integer $n \geq 3$, certain alternating maps of rank $n - 1$ have the property that $\mathrm{Pf}_{n-1}(f)$ is a Gorenstein ideal and that every Gorenstein ideal of height three arises in this way.

Let $\alpha = (a_{ij})$ be an $n \times n$ alternating matrix. $\mathrm{Pf}(\alpha)$ can be computed by development along any row as follows. For $1 \leq i \neq j \leq n$, let α_{ij} be the matrix obtained from α by deleting the ith and j^{th} rows and columns. Then for any i, $1 \leq i \leq n$,

$$\mathrm{Pf}(\alpha) = \sum_{j=1}^{n} a_{ij} C_{ij},$$

where, $C_{ij} = (-1)^{i+j-1}\mathrm{Pf}(\alpha_{ij})$, if $i < j$,

$C_{ij} = (-1)^{i+j}\mathrm{Pf}(\alpha_{ij})$, if $i > j$, and

$C_{ij} = 0$, if $i = j$.

Note that if α has a repeated row, then $\mathrm{Pf}(\alpha) = 0$.

We will need for the proof of (1.5), the fact that if F is a free R-module of odd rank n and $f\colon F \to F^*$ is an alternating map of rank $\leq n - 1$, then

$$\mathcal{I}_{n-1}(f) = (Pf_{n-1}(f))^2,$$

where $\mathcal{I}_{n-1}(f)$ is the ideal generated by the minors of [f] of order $n - 1$.

We can now prove the first part of the Buchsbaum-Eisenbud characterization which follows.

1.5. Theorem. Let $(R,\underline{m})$ be a regular local ring.

(1). Let $n \geq 3$ be an odd integer, and let F be a free R-module of rank n. Let $f: F \to F^*$ be an alternating map of rank $n - 1$ whose image is contained in $\underline{m}F^*$. If $Pf_{n-1}(f)$ has depth 3, then $Pf_{n-1}(f)$ is Gorenstein, and the minimal number of generators of $Pf_{n-1}(f)$ is n.

(2). Every Gorenstein ideal of R of height 3 arises as in (1).

Remark. Buchsbaum and Eisenbud's theorem is more general than (1.5). They define Gorenstein ideals appropriately for any local ring and then prove the structure theorem above for Gorenstein ideals of depth three with no hypothesis on the local ring $(R,\underline{m})$.

Proof of (1.5)(1). Fix a basis for F and corresponding dual basis for F^*. Let α be the corresponding matrix for f. Let γ_i be the Pfaffian of the alternating matrix obtained from α by deleting row i and column i. Let $g: R \to F$ be the map whose matrix with respect to the chosen basis of F is $\gamma = \mathrm{col}(\gamma_1, -\gamma_2, \gamma_3, -\gamma_4, \ldots, \gamma_n)$. Then the ith entry of $\alpha\gamma$ is the Pfaffian of an $n + 1 \times n + 1$ matrix obtained from α by repeating row i and column i. Hence $\alpha\gamma = 0$. Since $\alpha^* = -\alpha$, we have $\gamma^*\alpha = -\gamma^*\alpha^* = -(\alpha\gamma)^* = 0$. Thus we have a complex:

$$0 \to R \xrightarrow{g} F \xrightarrow{f} F^* \xrightarrow{g^*} R.$$

We will prove that this complex is exact using (Chapter 1; 6.5). Since coker $g^* = R/Pf_{n-1}(f)$, this will show that $Pf_{n-1}(f)$ is a Gorenstein ideal (by criterion (Chapter 1; 5.2).)

We first check the condition of (Chapter 1; 6.5) on the ranks of the maps g and f. We have assumed that f has rank $n - 1$. To show that g has rank 1 we need to show that some $\gamma_i \neq 0$. But, if we denote by $\mathscr{I}_j(\alpha)$, resp. $\mathscr{I}_j(g)$, the ideal of R generated by the minors of α (resp. of g) of order j, then $\mathscr{I}_{n-1}(\alpha) \neq 0$. But $(Pf_{n-1}(f))^2 = \mathscr{I}_{n-1}(\alpha) \neq 0$, so some $\gamma_i \neq 0$. To check the second condition of (Chapter 1; 6.5), we note that $(Pf_{n-1}(f))^2 = \mathscr{I}_{n-1}(\alpha)$ and $\mathscr{I}_1(g) = \mathscr{I}_1(g^*) = Pf_{n-1}(f)$. Thus the assumption that $Pf_{n-1}(f)$ contains an R-sequence of length three implies that the same is true for $\mathscr{I}_{n-1}(\alpha)$, $\mathscr{I}_1(g)$ and $\mathscr{I}_1(g^*)$.

Before we can give the proof of (1.5)(2), more spadework must be done. Now, suppose that I is a Gorenstein ideal of height three. Then R/I has, by (Chapter 1; 5.2), a minimal free resolution of the form

$$0 \to R \to F_2 \xrightarrow{f} F_1 \to R.$$

The idea behind Buchsbaum and Eisenbud's proof of (1.5)(2) is to define a multiplication on this resolution in such a way that it becomes a differential graded algebra and F_1 can be identified with F_2^*. Then they prove that f is alternating and that $I = Pf_{n-1}(f)$.

Let R be a commutative ring and let

$$\mathbb{F}: \quad \cdots \to F_3 \to F_2 \to F_1 \to R$$

be a free resolution with $F_0 = R$. Define the symmetric square $S^2(\mathbb{F})$ to be $S^2(\mathbb{F}) = \mathbb{F} \otimes \mathbb{F}/M$, where M is the graded submodule of $\mathbb{F} \otimes \mathbb{F}$ generated by $\{a \otimes b - (-1)^{(\deg a)(\deg b)} b \otimes a \mid a, b$ homogeneous elements of $\mathbb{F}\}$. The differential on $\mathbb{F} \otimes \mathbb{F}$ is defined by

$$d(a \otimes b) = da \otimes b + (-1)^{\deg a} a \otimes db.$$

Since $d(M) \subseteq M$, $S^2(\mathbb{F})$ is a quotient complex of $\mathbb{F} \otimes \mathbb{F}$. In degree ℓ we have:

$$S^2(\mathbb{F})_\ell \cong \coprod_{\substack{i+j=\ell \\ i<j}} F_i \otimes F_j, \text{ if } \ell \text{ is odd}$$

$$S^2(\mathbb{F})_\ell \cong \coprod_{\substack{i+j=\ell \\ i<j}} F_i \otimes F_j \oplus \Lambda^2 F_{\ell/2}, \text{ if } \ell/2 \text{ is an odd integer}$$

$$S^2(\mathbb{F})_\ell \cong \coprod_{\substack{i+j \\ i<j}} F_i \otimes F_j \oplus S^2 F_{\ell/2}, \text{ if } \ell/2 \text{ is an even integer.}$$

Thus $S^2(\mathbb{F})$ is a complex of free R-modules which is isomorphic to $\mathbb{F}$ in degrees 0 and 1. By the comparison theorem, there is a map of complexes φ: $S^2(\mathbb{F}) \to \mathbb{F}$ which extends the identification in degrees 0 and 1. We choose the map φ so that the restriction of φ to $R \otimes F_\ell \subseteq S^2(\mathbb{F})_\ell$ is the canonical isomorphism $R \otimes F_\ell \to F_\ell$. We define multiplication on F as follows:

$$a \cdot b = \varphi(\overline{a \otimes b}),$$

where $a, b \in \mathbb{F}$ and $\overline{a \otimes b}$ is the image of $a \otimes b$ in $S^2(\mathbb{F})$. The conditions that the map φ must satisfy are equivalent to the statement that $\cdot$ makes F into a (non-associative) graded, strictly skew-commutative differential algebra, with differential d, and with structure map $R \to \mathbb{F}$ given by the inclusion into degree 0.

Now we specialize to the case where $(R, \underline{m})$ is a regular local ring and I is a Gorenstein ideal of height n. Let

$$\mathbb{F}: \quad 0 \to F_n \xrightarrow{d_n} F_{n-1} \to \cdots \to F_1 \xrightarrow{d_1} R$$

be a minimal free resolution of R/I equipped with multiplication $\cdot$: $\mathbb{F} \otimes \mathbb{F} \to \mathbb{F}$ as above. Since I is Gorenstein, we may, by (Chapter 1; 5.2), identify F_n with R, so that for each $\ell \leq n$, the map

$$\cdot : F_\ell \otimes F_{n-\ell} \to F_n = R,$$

induces a map $s_\ell : F_\ell \to F^*_{n-\ell}$.

1.6. Proposition. For all $\ell \leq n$, s_ℓ is an isomorphism.

Proof. $\mathbb{F}$ and $\mathbb{F}^*$ are both minimal free resolutions of R/I. Moreover, we have the following diagram:

$$(\#) \quad \begin{array}{lccccccccccc} \mathbb{F}: & 0 \to F_n & \xrightarrow{d_n} & F_{n-1} & \to \cdots \to & F_2 & \xrightarrow{d_2} & F_1 & \xrightarrow{d_1} & R \\ & \| & & \downarrow s_{n-1} & & \downarrow s_2 & & \downarrow s_1 & & \| \\ \mathbb{F}^*: & 0 \to R & \xrightarrow{d_1^*} & F_1^* & \to \cdots \to & F_{n-2}^* & \xrightarrow{d_{n-1}^*} & F_{n-1}^* & \to & F_n^*. \end{array}$$

If we show that $s = (s_\ell)$ lifts the identification in degree 0, then all the s_ℓ must be isomorphisms by minimality. To do this we show that the diagram (#) commutes up to sign. If $a \in F_\ell$ and $b \in F_{n-\ell+1}$, then $a.b \in F_{n+1} = 0$, so

$$0 = d_{n+1}(a \cdot b) = d_\ell(a) \cdot b + (-1)^\ell f \cdot d_{n-\ell+1}(b),$$

and thus, $d_\ell(a) \cdot b = \pm a \cdot d_{n-\ell+1}(b)$. Consequently,

$$s_{\ell-1} d_\ell(a)(b) = d_\ell(a) \cdot b = \pm a \cdot d_{n-\ell+1}(b) = \pm s_\ell(a)(d_{n-\ell+1}(b))$$
$$= \pm\, d^*_{n-\ell+1}(s_\ell(a))(b).$$

Now we complete the proof of (1.5).

Proof of (1.5)(2). Assume that I is a Gorenstein ideal of height 3. Let

$$\mathbb{F}: \ 0 \to R \xrightarrow{f_3} F_2 \xrightarrow{f_2} F_1 \xrightarrow{f_1} R$$

be a minimal resolution of R/I. Then, $\mathbb{F}$ has the structure of a differential graded strictly skew-commutative algebra. By (1.6), the multiplication yields an identification of F_1 with F_2^*. We show that with this identification f_2 is alternating.

Let $\{e_i\}$, $1 \leq i \leq n$ be a basis of F_2 and $\{\varepsilon_i\}$, $1 \leq i \leq n$, be the dual basis of F_2^*, and let $\alpha = [a_{ij}]$ be the corresponding matrix of f_2. We want to show that $a_{ij} = -a_{ji}$ and that $a_{ii} = 0$. Now $a_{ij} = (f_2(e_j))(e_i)$, where $f_2(e_j)$ is regarded as a functional on F_2. We have then, $a_{ij} = (f_2(e_j))(e_i) = f_2(e_j) \cdot e_i \in R$. Now apply f_3 and the formula for differentiation of a product to obtain

$$f_3(a_{ij}) = f_3 f_2(e_j) \cdot e_i - f_2(e_j) \cdot f_2(e_i)$$

$$= -f_2(e_j) \cdot f_2(e_i) = -f_3(a_{ji}).$$

Similarly, one can show that $f_3(a_{ii}) = 0$. Since f_3 is a monomorphism, $a_{ij} = -a_{ji}$ and $a_{ii} = 0$.

We now show that $Pf_{n-1}(f_2) = I$. By the exactness criterion (Chapter 1; 6.5), rank $f_2 = n - 1$. Now, $\mathscr{I}(f_1)$ contains an R-sequence of length 3, so by (Chapter 1; 6.7), the same is true for $\mathscr{I}(f_2)$. Thus f_2 satisfies all the conditions of the first part of the theorem, (1.5)(1): we can, as in the proof of (1.5)(1), construct a map g; $R \to F_2$ so that

$$0 \to R \xrightarrow{g} F_2 \xrightarrow{f_2} F_2^* \xrightarrow{g^*} R$$

is exact and $\mathscr{I}(g) = Pf_{n-1}(f_2)$. Since both g and $f_3 = f_1^*$ are kernels of f_2, there must be a unit u in R such that $f_1^* = gu$. Then, $f_1 = ug^*$ and $I = Pf_{n-1}(f_2)$.

1.7. <u>Corollary</u>. Let I be a Gorenstein ideal of height 3 in a regular local ring $(R,\underline{m})$, then the minimal number of generators of I is odd.

<u>Example</u>. In [14] Buchsbaum and Eisenbud give the following examples of height 3 Gorenstein ideals in $R = k[[X,Y,Z]]$, k a field, with minimal number of generators any odd integer $n \geq 3$. Let H_n be the $n \times n$ matrix of the form

$$
H_n = \begin{bmatrix}
0 & X & 0 & \cdots\cdots & & 0 & Z \\
-X & 0 & Y & \cdots\cdots & & Z & 0 \\
0 & -Y & 0 & \cdots\cdots & & 0 & 0 \\
 & & & -Z \; 0 \; Z & & & \\
 & & & & & X & 0 \\
 & -Z & & & -X & 0 & Y \\
-Z & 0 .. & & & 0 & -Y & 0
\end{bmatrix}
$$

If we write $n = 2e + 1$, then X^e, Y^e and $Z^e \in Pf_{n-1}(H_n)$. Thus $Pf_{n-1}(H_n)$ has height three. Then (1.5)(1) applies to give that $Pf_{n-1}(H_n)$ is a Gorenstein ideal minimally generated by n elements.

It is conjectured that there are Gorenstein ideals of height 4 minimally generated by any number of generators ≥ 6. That 5 is not allowed follows from the result due to Buchsbaum and Eisenbud [15] and to Kunz [41] that a Gorenstein ideal of height h cannot be minimally generated by $h + 1$ elements.

1.8. Lemma. Let $(R,\underline{m})$ be a local ring and I an ideal. Let $\underline{x} = x_1,\ldots,x_n$ be a minimal set of generators and let $\underline{K} = K(\underline{x})$ be the associated Koszul complex. Assume that $H_1(\underline{K}) \cong R/I$ and that $H_j(\underline{K}) = 0$, for $j > 1$. Then R/I does not have finite projective dimension.

Proof. As suggested in [41] we will use Tate's process, cf. [67], of adjoining a variable to kill a cycle. We have $H_1(\underline{K}) = R/I\bar{\alpha}$, where $\bar{\alpha}$ is the homology class of a cycle α in $\underline{K}_1$. We construct a complex $\underline{\tilde{K}}$ as follows. Let $\underline{\tilde{K}}$ be the free $\underline{K}$-module on the basis $\{1,T,T^{(2)},T^{(3)},\ldots\}$ so that $\underline{\tilde{K}} = \underline{K} \oplus \underline{K}\,T \oplus \underline{K}\,T^{(2)} \oplus \cdots$. We grade $\underline{\tilde{K}}$ by giving $T^{(i)}$ the degree $2i$. Thus $\underline{\tilde{K}}_j = \coprod_{i=0}^{j} \underline{K}_{j-2i}T^{(i)}$. We extend the differential d on $\underline{K}$ to $\underline{\tilde{K}}$ by

$$dT^{(i)} = \alpha T^{(i-1)},$$

where $T^{(0)} = 1$ and $T^{(1)} = T$. To complete the proof, we will show that $\underline{K}$ is a minimal free resolution of R/I. Minimality follows from the fact that $\mathfrak{a} \in \underline{m}K$. To show that $\underline{K}$ is acyclic, let g be the R-linear map of complexes $\tilde{\underline{K}} \to \tilde{\underline{K}}$ defined by

$$g(c_0 + c_1T + \cdots + c_nT^{(n)}) = c_1 + c_2T + \cdots + c_nT^{(n-1)}.$$

g is of degree -2. The sequence

$$0 \to \underline{K} \xrightarrow{i} \tilde{\underline{K}} \xrightarrow{g} \tilde{\underline{K}} \to 0,$$

where i is the inclusion map, is exact so we get the long exact sequence of homology modules:

$$\to H_j(\underline{K}) \to H_j(\tilde{\underline{K}}) \to H_{j-2}(\tilde{\underline{K}}) \to H_{j-1}(\underline{K}) \to \cdots$$

$$\to H_3(\tilde{\underline{K}}) \to H_1(\tilde{\underline{K}}) \to H_2(\underline{K}) \to H_2(\tilde{\underline{K}}) \to H_0(\tilde{\underline{K}}) \to H_1(\underline{K}) \to H_1(\tilde{\underline{K}}) \to 0$$

But $H_i(\underline{K}) = 0$ for $i > 1$ and $H_0(\tilde{\underline{K}}) = H_0(\underline{K}) \cong R/I$. Since we killed $\mathfrak{a}$, $H_1(\tilde{\underline{K}}) = 0$. Now it follows easily, that $H_i(\tilde{\underline{K}}) = 0$ for $i > 0$.

1.9. Theorem. Let $(R,\underline{m})$ be a regular local ring and I a Gorenstein ideal of height h. Then I cannot be minimally generated by h + 1 elements.

Proof. Assume the contrary. We apply (Chapter 1; 4.2) to obtain that $H_1(\underline{K})$ is the highest non-vanishing homology of the Koszul complex associated with a fixed set of h + 1 generators for I, and that $H_1(\underline{K}) \cong \mathrm{Ext}_R^h(R/I,R)$. By (Chapter 1; 5.2), $\mathrm{Ext}_R^h(R/I,I) \cong R/I$. By (1.8), R/I does not have finite projective dimension which contradicts the fact that R is a regular local ring.

2. Cohen-Macaulay ideals.

2.1. Definition. Let $(R,\underline{m})$ be a local ring. An ideal I is a Cohen-Macaulay ideal (CM ideal, for short) if R/I is a Cohen-Macaulay ring.

Examples. If $(R,\underline{m})$ is a local ring, then all $\underline{m}$-primary ideals and all submaximal prime ideals are CM ideals. Thus, in particular, Macaulay's and Moh's primes are CM.

We look for some kind of a bound on the number of generators of CM ideals. We look first at the $\underline{m}$-primary ideals. Of course, if the dimension of the ring is greater than one, the number of generators of powers of $\underline{m}$ gets arbitrarily large, but it is natural to look for a bound for the number of generators of any $\underline{m}$-primary ideal I in terms of the least power t of $\underline{m}$ contained in I. We call t the nilpotency degree of R/I.

The following theorem given in [58] generalizes a result of Abhyankar [1].

2.2. Theorem. Let $(R,\underline{m})$ be a Cohen-Macaulay local ring of dimension $d > 0$. Let I be an $\underline{m}$-primary ideal and t the nilpotency degree of R/I. Then

$$v(I) \leq t^{d-1}e(R) + d - 1.$$

Proof. The proof is by induction on d. If $d = 1$, (2.2) follows from (Chapter 3; 1.1).

Assume that $d > 1$. Again, assuming that $R/\underline{m}$ is infinite as we may, there is a nonzero divisor x such that x is a superficial element for R. Pass to the $d - 1$ dimensional local Cohen-Macaulay ring R/x^tR. I/x^tR is $\underline{m}/x^tR$-primary so, by induction,

$$v(I/x^tR) \leq t^{d-2}e(R/x^tR) + d - 2.$$

Hence, $v(I) \leq v(I/x^tR) + 1 = t^{d-2}te(R) + d - 1$.

Remarks. (1). If $(R,\underline{m})$ is regular local and $d \geq 2$, then

$$v(I) \leq gt^{d-2} + d - 1,$$

where g is the degree of I, i.e., $I \subseteq \underline{m}^g \setminus \underline{m}^{g+1}$.

(2). In [1], Abhyankar shows that the CM hypothesis is necessary in (2.2) for $I = \underline{m}$.

(3). The rings $k[[t^n, t^{n+1}, \ldots, t^{2n-1}]]$, k a field, provide examples of maximal ideals where the bound in (2.2) is attained.

Now we use the result for $\underline{m}$-primary ideals to obtain a bound for CM ideals of arbitrary height.

2.3. Theorem. Let $(R,\underline{m})$ be a d dimensional local Cohen-Macaulay ring. Let I be a Cohen-Macaulay ideal of height $h > 0$. Then

$$v(I) \leq e(R/I)^{h-1} e(R) + h - 1.$$

Proof. We may assume that $R/\underline{m}$ is infinite. The proof is by induction on $s = \dim(R/I)$. If $s = 0$ then by (2.2) it is sufficient to note that $e(R/I) = \lambda(R/I) \geq$ nilpotency degree of R/I.

Assume $s > 0$. By (Chapter 1; 3.2), there is a nonzero divisor x in $\underline{m}$ such that x is a superficial element for R and the image of x in R/I is a nonzero divisor in R/I and a superficial element for R/I. We pass to the $d - 1$ dimensional Cohen-Macaulay ring R/xR and to the height h Cohen-Macaulay ideal $(I,x)/xR$. By induction,

$$\begin{aligned} v(I) &= v((I,x)/xR) \leq e(R/(I,x))^{h-1} e(R/xR) + h - 1 \\ &= e(R/I)^{h-1} e(R) + h - 1. \end{aligned}$$

Remarks and Examples.

(1). For height 1 ideals I, (2.3) gives Rees' theorem [56]. If height $I = 2$, Rees [56] has the result, $v(I) \leq e(R) + e(R/I)$, which gives a better bound than (2.3) except when R or R/I is regular when they coincide.

(2). If R is a regular local ring and I is a CM ideal, then

$$v(I) \leq e(R/I)^{h-1} + h - 1,$$

with equality if and only if R/I is regular. To see this, note that if I is $\underline{m}$-primary we have by (2.2), $v(I) \leq t^{d-1} + d - 1$. It is clear that $t < e(R/I)$ except when $\underline{m}/I$ is principal in which case $v(I) = d$ and $e(R/I) > 1$ unless $I = \underline{m}$. If I is not $\underline{m}$-primary, then pick x in $\underline{m} \setminus \underline{m}^2$ as in the proof of (2.3) and note that R/xR is regular. A non-Gorenstein prime P of height 2 and $e(R/P) = 3$ provides an example where $v(P) = e(R/I)^{h-1} + h - 2$. For example, take P to be the kernel of the homomorphism $k[[X,Y,Z]] \to k[[t^3,t^4,t^5]]$.

3. Weakening the hypotheses on I and R.

We would like to have bound for numbers of generators of ideals which are not quite CM, eg. for ideals I in R regular such that R/I is 3 dimensional normal. Results in this direction have been proved by Becker [8] and Hartshorne and Ogus [31] with additional hypotheses on R, eg., R is equicharacteristic.

Before examining some results of this type, we show that some vestige of the CM hypothesis must remain. First of all, for there to be any hope of some relation between $v(I)$ and $e(R/I)$, I must certainly be unmixed as the following example shows. Let $R = k[[X,Y]]$, k a field. Let $I_n = (X^n, X^{n-1}Y, \ldots, XY)$. Then $v(I_n) = n$ and $e(R/I) = 1$.

Much more subtle is the following example of Hochster [8]. He has constructed primes P_n of height 4 in a polynomial ring T in 7 variables over the complex numbers $\mathbb{C}$ such that $v(P_n) \geq n$ and $e(T/P_n) = 2$.

Example. Let $R = \mathbb{C}[[X,Y,Z]]$. First note that for each integer $n > 0$, there exists a_n, b_n, c_n in $(X,Y,Z)R$ such that if L_n is the kernel of the homomorphism $R^3 \twoheadrightarrow (a_n, b_n, c_n)$ then L_n needs at least n generators. Let Q_n be primes of height 2 in R such that Q_n needs at least n generators. Take a_n, b_n an R-sequence of length 2 in Q_n and pick c_n so that $Q_n = ((a_n, b_n) : c_n R)$. Then Q_n is a homomorphic image of L_n via projection on the last factor, so L_n needs at least n generators since Q_n does.

Let $S_n = \mathbb{C}[[X,Y,Z,W^2,a_nW,b_nW,c_nW]]$. Then $R \subset S_n \subset \mathbb{C}[[X,Y,Z,W]]$ and $e(S_n) = 2$, cf. (Chapter 1; 3.5).

Define a homomorphism $\psi_n: R[V_4,V_5,V_6,V_7] \to S_n$ by $\psi_n(V_4) = W^2$, $\psi_n(V_5) = a_nW$, $\psi_n(V_6) = b_nW$, $\psi_n(V_7) = c_nW$ and ψ_n is the identity on R. We will show that $P_n = \ker \psi_n$ requires at least n generators.

Consider the following map of graded R-algebras:

$$\varphi_n: R[V_4,V_5,V_6,V_7] \to R[W]$$

defined by: φ_n is the identity on R, $\varphi_n(V_4) = W^2$, $\varphi_n(V_5) = a_nW$, $\varphi_n(V_6) = b_nW$ and $\varphi_n(V_7) = c_nW$, where the variables all have degree 1 except V_4 has degree 2. φ_n preserves degrees, so $\ker \varphi_n$ is a homogeneous ideal of $R[V_4,V_5,V_6,V_7]$. $\ker \varphi_n$ contains no elements of degree 0. An element of degree 1 in $\ker \varphi_n$ corresponds to a triple of elements (α,β,γ) in R such that $\alpha a_n + \beta b_n + \gamma c_n = 0$. Thus it follows that a minimal homogeneous basis for $\ker \varphi_n$ must contain a basis for L_n, the kernel of the map $R^3 \to (a_n,b_n,c_n)$, i.e., $\ker \varphi_n$ requires at least n generators. Now ψ_n arises from φ_n by completion, so it follows that $\ker \psi_n$ requires at least n generators.

<u>Note</u>. In $C[[X,Y]]$ ideals of the form $((a_n,b_n) : c_n)$ are all generated by 2 elements.

If I is a "little less than" CM and if R is an equicharacteristic CM local ring then there is a bound for $v(I)$ in terms of $e(R/I)$. The following theorem is a slight generalization to CM rings of Becker's result [8] for regular rings.

3.1. <u>Theorem</u>. Let $(R,\underline{m})$ be an equicharacteristic Cohen-

Macaulay local ring of dimension d and I an ideal of height h such that $I\hat{R}$ is unmixed and depth $R/I \geq \dim R/I - 1$. Then

$$v(I) \leq e(R/I)^h e(R) + h - 1.$$

Proof. We may assume that R is complete and that I is an unmixed ideal with depth $R/I \geq \dim R/I - 1$. We may also assume that $R/\underline{m}$ is infinite. Let k be a coefficient field in R. Pick elements $x_1,\ldots,x_r,\ldots,x_d$ in R such that R is a free module of rank $e(R)$ over $k[[x_1,\ldots,x_d]]$ and such that $\overline{x}_1,\ldots,\overline{x}_r$, where $\overline{x}_i$ is the image of x_i in R/I, is a system of parameters in R/I with the property that $e(R/I) = e((\overline{x}_1,\ldots,\overline{x}_r))$, cf. (Chapter 1; 3.3). Since $\overline{x}_1,\ldots,\overline{x}_r$ are analytic indeterminates over k we identify $A = k[[x_1,\ldots,x_r]]$ with $k[[\overline{x}_1,\ldots,\overline{x}_r]]$. By [70; Remark, p. 293], R/I is a finitely generated A-module. For $j = 1,\ldots,d - r$, let f_{r+j} be a monic polynomial over A in x_{r+j} of minimal degree δ_{r+j} satisfied by $\overline{x}_{r+j}$. Let $J_0 = (f_{r+1},\ldots,f_d)k[[x_1,\ldots,x_d]]$. Let $J = J_0R$. We have then

$$A \subset k[[x_1,\ldots,x_d]]/J_0 \subset R/J.$$

$k[[x_1,\ldots,x_d]]/J_0$ is a free A-module of rank $\delta_{r+1} \cdots \delta_d$, consequently, R/J is a free A-module of rank $e(R)\delta_{r+1} \cdots \delta_d$. By (Chapter 1; 3.5), $e((x_1,\ldots,x_r)R/J) = e(R)\delta_{r+1} \cdots \delta_d$.

Now consider the exact sequence of A-modules

$$0 \to I/J \to R/J \to R/I \to 0.$$

It follows from (Chapter 1; 5.5) and our hypothesis on depth R/I, that R/I has projective dimension at most one over A and thus, that I/J is a free A-module. This means that the exact sequence of A-modules

$$0 \to J \to I \to I/J \to 0$$

splits, and therefore, that $v(I) \leq v(J) + rk_A(I/J)$. Since the hypothesis on I implies that R/I is torsion free over A, we have

$$v(I) \leq d - r + e(R)\delta_{r+1} \cdots \delta_d - 1 = e(R)\delta_{r+1} \cdots \delta_d + h - 1.$$

It remains to show that $\delta_{r+j} \leq e(R/I)$, for $j = 1,\ldots,d - r$. But, by (Chapter 1; 3.5), $e(R/I)$ is the maximum number of linearly independent elements in R/I over A, i.e., the dimension of the total quotient ring of R/I over the quotient field Q of A. Thus the minimal polynomial for $\overline{x}_{r+j}$ over Q has degree at most $e(R/I)$. But A integrally closed implies that this minimal polynomial has coefficients in A. Hence $\delta_{r+j} \leq e(R/I)$.

Remark. That the exponent of $e(R/I)$ in (3.1) cannot be reduced to $h - 1$ as in (2.3) is shown by the example $R = k[[X,Y,Z]]$ with $P = (X^3-Z^2, XY^2-W^2, XW-YZ, X^2Y-ZW)$. P is a height 2 prime, $R/P \cong k[[U^2,V,U^3,UV]]$ is not CM but depth $R/P = \dim R/P - 1$. However, $v(P) = 4$ and $e(R/P) = 2$.

As a corollary of the two dimensional theorem (Chapter 3; 2.6) we have the following result for certain "not quite CM" ideals in a CM local ring of arbitrary dimension.

3.2. Theorem. Let $(R,\underline{m})$ be a Cohen-Macaulay local ring of dimension d. Let I be an ideal with depth $R/I \geq d - 2$. Then

$$v(I) \leq (n + 1)e(R),$$

where $n = \dim_{R/\underline{m}} \mathrm{Ext}_R^{d-2}(R/\underline{m}, R/I)$.

Proof. Assume that $R/\underline{m}$ is infinite and let $x_1,\ldots,x_d$ be elements of R such that $e(R) = e((x_1,\ldots,x_d)) = e(R/(x_1,\ldots,x_i)R)$, for $1 \leq i \leq d$, and such that the images of $x_1,\ldots,x_{d-2}$ in R/I form an R/I sequence. Then $v(I) = v((I,x_1,\ldots,x_{d-2})/(x_1,\ldots,x_{d-2}))$, $e(R) = e(R/(x_1,\ldots,x_{d-2}))$ and $\mathrm{Ext}_R^{d-2}(R/\underline{m},R/I) \cong$

$\mathrm{Hom}_{R/(x_1,\ldots,x_{d-2})}(R/\underline{m}, R/(x_1,\ldots,x_{d-2}))$. Thus we have reduced the problem to (Chapter 3; 2.6).

(3.2) gives us another way of obtaining Rees' result [56]:

3.3. Corollary. Let $(R,\underline{m})$ be a local Cohen-Macaulay ring of dimension d. Let I be a height 1 Cohen-Macaulay ring of dimension d. Let I be a height 1 Cohen-Macaulay ideal. Then

$$v(I) \leq e(R).$$

Proof. Since R/I is CM, $\mathrm{Ext}_R^{d-2}(R/\underline{m}, R/I) = 0$.

For other results of this type, obtained using entirely different methods, the reader should consult the paper [31] of Hartshorne and Ogus. Using duality theory, they have results on complete intersections where I is only assumed to be a "half-way" CM ideal, i.e., depth $R/I \geq \frac{1}{2}(\dim R/I + 1)$.

REFERENCES

1. S. S. Abhyankar, Local rings of high embedding dimension, Amer. J. Math. 89 (1967), 1073-1077.
2. S. S. Abhyankar, On Macaulay's examples, Conference on Commutative Algebra, Lect. Notes in Math. 311, Springer-Verlag, Berlin, 1972.
3. E. Artin, Geometric Algebra, Interscience, New York, 1957.
4. Y. Akizuki, Zur Idealtheorie der einartigen Ringbereiche mit dem Teilerkettensatz, Jap. J. Math. 14(1938), 85-102.
5. M. Auslander and D. A. Buchsbaum, Codimension and multiplicity, Ann. Math. 68 (1958), 625-657.
6. M. Auslander and D. A. Buchsbaum, Homological dimension in local rings, Trans. Amer. Math. Soc. 85(1957), 390-405.
7. H. Bass, On the ubiquity of Gorenstein rings, Math. Zeit. 82 (1963), 8-28.
8. J. Becker, On the boundedness and unboundedness of the number of generators of ideals and multiplicity, preprint.
9. M. Boratynski and D. Eisenbud, On the number of generators of ideals in local Macaulay rings of dimension two, preprint.
10. M. Boratynski and J. Swiecicka, The Hilbert-Samuel function of a Cohen-Macaulay ring, preprint.
11. N. Bourbaki, Algèbre Commutative, Actualités Scientifiques et Industrielles, Hermann, Paris, 1961-1965.
12. H. Brezinsky, On prime ideals with generic zero $x_i = t^{n_i}$, Proc. Amer. Math. Soc. 47 (1975), 329-332.
13. D. A. Buchsbaum, Lectures on regular local rings, Category Theory, Homology Theory and Their Applications I, Lecture Notes in Math. 86, Springer-Verlag, Berlin, 1969.
14. D. A. Buchsbaum and D. Eisenbud, Algebra structures for resolutions and some structure theorems for ideals of codimension 3, Am. J. Math., to appear.
15. D. A. Buchsbaum and D. Eisenbud, Remarks on ideals and resolutions, Symp. Math. XI (1973), 191-204.

16. D. A. Buchsbaum and D. Eisenbud, Some structure theorems for finite free resolutions, Adv. in Math. 18(1975), 245-301.

17. D. A. Buchsbaum and D. Eisenbud, What makes a complex exact? J. of Alg. 25 (1973), 259-268.

18. L. Burch, Codimension and analytic spread, Proc. Camb. Phil. Soc. 72 (1972), 369-373.

19. H. Cartan and S. Eilenberg, Homological Algebra, Princeton Mathematical Series 19, Princeton University, 1956.

20. S. U. Chase, Direct products of modules, Trans. Amer. Math. Soc. 97 (1960), 457-473.

21. C. Chevalley, On the theory of local rings, Ann. Math. 44 (1943), 690-708.

22. I. S. Cohen, Commutative rings with restricted minimum condition, Duke Math. J. 17 (1950), 27-42.

23. E. D. Davis, Regular sequences and minimal bases, Pac. J. Math. 26 (1971), 323-326.

24. J. A. Eagon and D. G. Northcott, Ideals defined by matrices and a certain complex associated with them, Proc. Roy. Soc. A 269 (1962), 188-204.

25. J. A. Eagon and D. G. Northcott, On the Bucksbaum-Eisenbud theory of finite free resolutions, J. Reine Angew. Math. 262/263 (1973), 205-219.

26. P. Eakin and A. Sathaye, Prestable ideals, J. of Alg. 41 (1976), 439-454.

27. D. Eisenbud, Subrings of Artinian and Noetherian rings, Math. Ann. 185 (1970), 247-249.

28. D. Ferrand, Suite régulière et intersection complete, C. R. Acad. Sc. Paris 264 (1967), 427-428.

29. W.-D. Geyer, On the number of equations which are necessary to describe an algebraic set in n-space, preprint.

30. A. Grothendieck, Eléments de Géométrie Algébrique IV (Seconde Partie), IHES, Paris, 1964.

31. R. Hartshorne and A. Ogus, On the factoriality of local rings of small embedding codimension, Comm. in Alg. 1 (1974), 415-437.

32. J. Herzog, Generators and relations of abelian semigroups and semigroup rings, Manu. Math. 3 (1970), 175-193.

33. J. Herzog and E. Kunz, Der kanonische Modul eines Cohen-Macaulay-Rings, Lect. Notes in Math. 238, Springer-Verlag, Berlin, 1971.

34. J. Herzog and R. Waldi, A note on the Hilbert function of a one-dimensional Cohen-Macaulay ring, Manu. Math. 16 (1975), 251-260.

35. M. Hochster, Topics in the Homological Theory of Modules over Commutative Rings, Regional Conference Series in Math. 24, Amer. Math. Soc., 1975.

36. M. Hochster and J. A. Eagon, Cohen-Macaulay rings, invariant theory and the generic perfection of determinantal loci, Amer. J. Math. 93 (1971), 1020-1050.

37. M. Hochster and L. J. Ratliff, Jr., Five theorems on Macaulay rings, Pac. J. Math. 44 (1973), 147-172.

38. M. Hochster and J. Roberts, Rings of invariants of reductive groups acting on regular rings are Cohen-Macaulay, Adv. in Math. 13 (1974), 115-175.

39. I. Kaplansky, Commutative Rings, Allyn and Bacon, Boston, 1970.

40. D. Kirby, The reduction number of a one-dimension local ring, J. Lon. Math. Soc. (2) 10 (1975), 471-481.

41. E. Kunz, Almost complete intersections are not Gorenstein rings, J. of Alg. 28 (1974), 111-115.

42. E. Kunz, The value semigroup of a one-dimensional Gorenstein ring, Proc. Amer. Math. Soc. 25 (1970), 748-751.

43. D. Laksov, The arithmetic Cohen-Macaulay character of Schubert schemes, Acta Math. 129 (1972), 1-9.

44. S. Lichtenbaum and M. Schlessinger, The cotangent complex of a morphism, Trans. Amer. Math. Soc. 128 (1967), 41-70.

45. J. Lipman, Stable ideals and Arf rings, Amer. J. Math. 93 (1971), 649-685.

46. F. S. Macaulay, The Algebraic Theory of Modular Systems, Cambridge University, 1916.

47. E. Matlis, The multiplicity and reduction number of a one dimensional local ring. Proc. Lon. Math. Soc. XXVI (1973), 273-288.
48. H. Matsumura, Commutative Algebra, W. A. Benjamin, New York, 1970.
49. N. McCoy, Rings and Ideals, The Carus Math. Monographs, 8, Math. Assoc. of Amer. 1948.
50. T. T. Moh, On the unboundedness of generators of prime ideals in power series rings of three variables, J. Math. Soc. Japan 26 (1974), 722-734.
51. M. Nagata, Local Rings, Interscience, New York, 1962.
52. D. G. Northcott and D. Rees, Reductions of ideals in local rings, Proc. Camb. Phil. Soc. 50 (1954), 145-158.
53. C. Peskine and L. Szpiro, Dimension projective finie et cohomologie locale, Publ. Math. I.H.E.S. 42, Paris, 1973, 323-395.
54. Y. Quentel, Sur l'uniforme cohérence des anneaux noethériens, C. R. Acad. Sc. Paris 275 (1972), 753-755.
55. M. Raynaud and L. Gruson, Critères de platitude et de projectivité, Inv. Math. 13 (1971), 1-89.
56. D. Rees, Estimates for the minimum number of generators of modules over Cohen-Macaulay local rings, preprint.
57. J. D. Sally, Boundedness in two dimensional local rings, Amer. J. Math., to appear.
58. J. D. Sally, Bounds for numbers of generators of Cohen-Macaulay ideals, Pac. J. Math. 63 (1976), 517-520.
59. J. D. Sally, On the associated graded ring of a local Cohen-Macaulay ring, Kyoto J. Math., to appear.
60. J. D. Sally and W. V. Vasconcelos, Stable rings, J. Pure and Appl. Alg. 4 (1974), 319-336.
61. J.-P. Serre, Algèbre Locale, Multiplicitiés, 3rd edition, Lect. Notes in Math. 11, Springer-Verlag, Berlin, 1975.
62. J.-P. Serre, Sur la dimension homologique des anneaux et des modules noetherian, Proc. Int. Symp., Tokyo-Nikko, 1955, Sc. Council of Japan, Tokyo, 1956, 175-189.
63. B. Singh, Effect of a permissible blowing-up on the local Hilbert function, Inv. Math. 26 (1974), 201-212.

64. J. P. Soublin, Anneaux et modules cohérents, J. of Alg. 15 (1970), 455-472.

65. R. P. Stanley, Hilbert functions of graded algebras, preprint.

66. L. Szpiro, Variétés de codimension 2 dans P^n, Colloque d'Algèbre de Rennes, 1972, exp. 15.

67. J. Tate, Homology of Noetherian rings and local rings, Ill. J. Math. 1 (1957), 14-27.

68. W. V. Vasconcelos, A note on normality and the module of differentials, Math. Zeit. 105 (1968), 291-293.

69. J. Watanabe, A note on Gorenstein rings of embedding codimension three, Nagoya Math. J. 50 (1973), 227-232.

70. O. Zariski and P. Samuel, Commutative Algebra II, D. Van Nostrand, Princeton, 1960.

INDEX

about the book . . .

The central theme of this book is the problem of determining the number of generators of an ideal in a local ring. The book gathers together and discusses in detail the major results on this subject which have appeared since 1965 and which use primarily algebraic techniques. In the process of gathering these results, the book examines the directions of ongoing research and pinpoints the central open problems in this area. To make these lecture notes as self-contained as possible, the author includes statements and proofs of all results which are used in the text but not found in the basic reference books on commutative algebra.

Numbers of Generators of Ideals in Local Rings makes an important contribution to the work of research mathematicians interested in local algebra, and is accessible to graduate students with a basic background in commutative algebra.

about the author . . .

JUDITH D. SALLY is Associate Professor in the Department of Mathematics at Northwestern University in Evanston, Illinois, where she has been teaching since 1972. Dr. Sally's research interests lie in the study of local rings with particular emphasis on questions concerning numbers of generators of ideals in local rings and Hilbert functions. Dr. Sally received her Ph.D. degree from the University of Chicago in 1971. She currently holds an Alfred P. Sloan Research Fellowship. She is a member of the American Mathematical Society and the Association for Women in Mathematics.

Printed in the United States of America ISBN: 0–8247–6645–8

marcel dekker, inc./new york · basel